Management Division

Presentations at the 2008 AIChE Annual Meeting

Philadelphia, Pennsylvania
16 - 21 November 2008

ISBN: 978-1-61567-220-2

Printed from e-media with permission by:

Curran Associates, Inc.
57 Morehouse Lane
Red Hook, NY 12571
www.proceedings.com

Some format issues inherent in the e-media version may also appear in this print version.

Printed by Curran Associates, Inc. (2009)

AIChE
3 Park Avenue
New York, NY 10016-5991

Phone: (203) 702-7660
Fax: (203) 775-5177

www.aiche.org

TABLE OF CONTENTS

Smart Management Strategies for Smart Manufacturing Plants

Jeremiah O'Brien
Process Data Control Corp
Arlington, Texas

Jimmy L. Humphrey Ph.D., P.E.
J. L. Humphrey & Associates
Austin, Texas

Introduction

This paper reviews the evaluation, management, and implementation aspects of Smart Plants, as elucidated by a "Smart Distillation" survey performed by Dr. Jimmy Humphrey in the fall of 2007. It asked the following questions:

Suppose all key distillation columns within a company were equipped so that accurate real time material and heat balances could be determined around each column, and further suppose that data may be shared locally and globally with the enterprise in real time -- what advantages do you see, and what are the challenges?

Why a "Smart Distillation" Survey?

"Smart Distillation" is one of many design and operational innovations that fall under the general category of "Smart Plants." The purpose of moving to "Smart Plant" technologies is to improve a facility's ability to predict problems and act proactively to prevent problems before they occur.

Globalization, availability of feedstocks and fuels, and the retirements of an aging workforce, are some of the challenges facing process industries today. It is an opportune time for the process industries, universities and government to collaborate on new programs to upgrade existing plants or build new plants based on to "Smart Plant" design principles.

Why choose distillation to explore this topic? Because even though extensive amounts of research have been done over the past 30 years on alternative processes, there is no replacement in sight for distillation. About 80,000 distillation columns are in operation in existing chemical plants and petroleum refineries worldwide, about one-half of which are in the U.S. These columns, with a thermodynamic efficiency of only 3%-5%, consume the energy equivalent of 2.5 to 3.0 million BPD of crude oil worldwide. Distillation is critical to process industries, contributing directly to revenues of about $6 trillion per year.

Why Evaluate "Smart Plant" Alternatives?

The issues that "Smart Plants" are intended to address are numerous and costly to U.S. process industries. The authors estimate that reductions in excessive energy consumption through "Smart Plant" innovations, for example, could save industries $20 billion per year. Avoidance of unplanned plant shutdowns could save an additional $20 billion per year. Improved targeting of maintenance, which would avoid both "too much too soon" and "too little too late," could save $5 billion per year. Because a single environmental or safety incident can cost a company as much as $2 billion, the cost savings from avoiding or reducing incidents is potentially larger than all of the other projected savings combined.

What We Learned

Approximately fifty individuals responded to the "Smart Distillation" survey. One-half of these responses came from individuals in process industries, and the other half came from individuals in universities, suppliers, government, and consulting firms. Responses included both direct answers to the questions and more general feedback and suggestions, as well as examples of pertinent industry initiatives. As a whole, the respondents confirmed that there is interest in "Smart Distillation" and "Smart Manufacturing" by process industries, operating companies, universities, and government agencies.

Many of the "Smart Distillation" advantages identified by respondents go beyond distillation to include unit operations, such as heat transfer, and other industries, such as telemedicine. Similarly, the key challenges identified by respondents covered a broad range of topics, such as processing large volumes of data in a timely manner. Both the advantages and key challenges are discussed in the next two sections.

Main Advantages of "Smart Distillation"

Respondents to this survey identified 70 different types of advantages that could result from the proposed instrumentation and data utilization that would results from "Smart Distillation." The principal "Smart Distillation" benefits identified by respondents included the following:

- Monitor, diagnose, and troubleshoot distillation columns (heat exchangers, furnaces, reactors, etc.) via broadband using centrally located, highly skilled technical staffs.
- Constantly compare performance of similar columns (globally) to identify and improve those that are underperforming.
- Perform process diagnostics on individual columns and compare actual with theoretical performance.
- Detect unexpected leaks earlier and save money by maintaining safe operations.
- Continuously operate columns at tighter product specifications to reduce off-spec products and increase energy efficiency.
- Having the energy requirements of each column, an online version of the Total Site Analysis could be determined (using Site Utility Curve).
- Constantly update distillation models and use them for accurate optimization.
- Optimize capital equipment to reduce carbon footprint of the global enterprise.
- Operate columns closer to their flood points to maximize production.

- Enable R&D and Process Improvement groups to mine data for validating process simulation theories and control strategies.
- Improve MHB awareness of operators and others to prevent safety incidents such as occurred recently in U.S.
- Compare performance of plants versus commercially available technology to make better capital investment decisions.
- Coordinate and schedule production from information that is available globally to maximize overall profits.
- Foster relationships among industry, university, and government to sponsor student interns to develop predictive models using accurate real time MHB.

Key Challenges to Implementing "Smart Distillation"

Respondents to this survey also identified several potential drawbacks and concerns about "Smart Distillation." The key challenges cited most often are:

- Is there such a thing as an accurate material and heat balance? Are compositions and flow rates based on inaccurate measurements? Can compositions be measured in real time? Distillation does not always operate as a steady state process;
- More emphasis needs to be placed on applying wired and wireless sensors as well as soft sensors. Getting accurate real time measurements, and applying accurate predictive models, are critical to taking proactive actions that prevent plant problems before they occur;
- How do we handle the volumes of data that will be generated? These data may cause confusion and delays without considerable supporting software and graphics. Can we develop intelligent data analyses to quickly process the amount of data generated and pinpoint the problems?
- How can benefits outweigh the costs for small plants? and
- Will we have enough people with the expertise to intelligently mine the data?

Where Industry Is Today

Some process industry companies are sharing plant data globally via the Internet to enable in-house "experts" access to this information. One respondent described an application that he is using to track the extent to which some unit operations vary from a theoretical optimum state, and receive financial reports that show dollars lost in real time. We also learned about a major chemical company that implemented a centralized monitoring system to achieve environmental management, which includes a dozen plants located in multiple states.

Some respondents reported that they are in the early stages of developing and sharing the benefits of real time data utilization. They are currently using:

- Broadband to access data historians to get real time plant data. These data are used for monitoring, diagnosing, troubleshooting, and modeling.
- Real time optimizers that use third-party software (e.g., AspenTech's RTOPT) on individual distillation units.

- Advanced control systems that preferentially feed the most efficient distillation columns first, to the extent demand permits.
- Systems to track how far off optimum some unit operations are and translating the difference into dollars lost in real time.
- Centralized monitoring system with web access for Title V Compliance and process historian data collection and evaluation via broadband

Wireless instrumentation has not made significant inroads in the process industries sector because most plants have not yet installed wireless networks outside of administrative offices and control rooms. However, wireless technology offers significant cost, versatility, and other advantages to manufacturing plants, which will enable it to dominate the instrumentation market in the coming years. When those conditions prevail, more instrumentation will be installed in operating units, opening the door to a number of "Smart Plant" innovations.

Knowledge Management Considerations

In a published work entitled "Position Paper on Knowledge Asset Management," Professor Ann Macintosh of the Artificial Intelligence Applications Institute (U. of Edinburgh) explained the importance of information "assets" to an organization and suggested four criteria for assessing Knowledge Management (KM). These criteria are explained below (items in italics were added by the authors):

1. Identify the knowledge assets a company possesses
    - Where is the knowledge asset?
    - What does it contain?
    - What is its use and what form is it in?
    - How accessible is it?

2. Analyze how the knowledge can add value ("Create Added Value")
    - What are the opportunities for using the knowledge asset?
    - What would be the effect of its use?
    - What are the current obstacles to its use?
    - What would be its increased value to the company?

3. Specify what actions are necessary to achieve better usability and benefit ("Make Knowledge Usable")
    - How to plan the actions to use the knowledge asset?
    - How to enact actions?
    - How to monitor actions?

4. Review the use of the knowledge to ensure added value ("Ensure Value is Added")
    - Did the use of it produce the desired added value?
    - How can the knowledge asset be maintained for this use?

- Did the use create new opportunities?

The chart below assesses each of identified Smart Distillation advantages against the last three Professor Macintosh KM criteria, A subjective rating scale of Smart Distillation advantages, from 1 (no effect) to 5 (high impact), was used to rank order the advantages based on how well KM criteria would be supported.

Smart Distillation Advantages (summarized)	Create Added Value	Make Know-ledge Usable	Ensure Value is Added	Total Score
Constantly update distillation models for accurate optimization.	4	4	3	11
Detect unexpected leaks earlier and save money by maintaining safe operations.	5	3	3	11
Improve coordination and scheduling of production to maximize profits.	3	4	4	11
Validate process simulation theories and control strategies.	3	4	4	11
Operate columns at tighter specs to reduce off-spec products & increase efficiency.	5	4	2	11
Compare performance to available technology to improve capital investments	3	4	3	10
Operate columns closer to their flood points to maximize production.	4	4	2	10
Optimize capital equipment to reduce carbon footprint	3	4	3	10
Compare performance of similar columns to improve underperforming. assets	4	3	2	9
Determine energy requirements of each column using Site Utility Curve	3	3	2	8
Perform process diagnostics to compare actual vs. theoretical performance.	3	3	2	8
Centrally Monitor, diagnose, and troubleshoot equipment via broadband	3	3	1	7
Improve operator awareness of MHB to prevent safety incidents	3	1	1	5
Sponsor student interns to develop predictive models using real time MHB.	2	1	1	4

KM is sufficiently important to warrant careful analysis of the potential advantages of various design and operation approaches on criteria such as the ones proposed by Dr. Macintosh. After ranking the potential advantages of a technological innovation, or comparing the KM impacts among competing ideas, planners and designers can move toward adopting approaches that are most likely to achieve optimal project results from a KM perspective.

Evaluation

In addition to assessing the advantages of "Smart Distillation" (and other Smart Plant technologies) against standard knowledge management (KM) criteria, planners must also consider life cycle cost, technical feasibility, work force acceptance, regulatory approvals, competing priorities, and numerous other factors. At the same time, one reason why plants have not always been designed or operated as "smart" as possible is that decision makers have often overlooked KM considerations.

How can KM be assured of playing a role in the design and operation of process industry plants? One way is to focus on the value-added component. Authors of process improvement books often advise readers to look at other fields of endeavor for insights into solving their problems. In this context, the Ohio Department of Education may offer a valuable perspective into assessing value-added considerations. The case in point involves measuring student progress.

The Ohio Accountability Task Force was created in House Bill 3 to guide the implementation of 'value-added' progress measures into the accountability system of the department. House Bill 3 mandated that the previously used "Performance Index Growth" method be abandoned in favor of a new value-added approach to be developed by the Task Force.

Task Force members soon realized that "Value-Added Measures" were needed to establish metrics for assessing progress because previous methods relied on metrics unrelated to the core values of the department. House Bill 3 demanded that the evaluations be based on metrics that took into account and accurately reflected the core values of the department related to student progress. Even relatively objective measures, such as achievement test scores, did not meet this test, because the results were not scaled to a core department value. Test scores were therefore scaled based on factors tied to the goals of each school and district, and so-called Composite Scores were then used to evaluate added value.

There are three lessons to be learned from this example. First, metrics are essential to measuring the value added from innovation; in other words, "value" must ultimately be quantifiable. Second, just because there are performance indicators that can be measured and compared to historical data, does not mean that they will properly evaluate value added. Third, and perhaps most importantly, metrics that reflect the core values and goals of an organization are of greatest benefit to decision-makers when evaluating alternatives based on value added.

Implementation

"Smart Distillation" will require the installation of new sensors, computer network components, and other improvements, in addition to the capacity to process a large amount of information. The challenges in meeting these requirements were pointed out by several respondents to the survey. This raises an implementation issue that is especially germane to "Smart Plants," namely, should decision makers wait until every part of an innovation is supported by proven and available infrastructure before proceeding? Or, should implementation be commenced even with some uncertainty and "leaps of faith?"

One of the author's experiences with technology for detecting changes to a company's applicable environmental rules may provide some insights into this question. About ten years ago, industries began to comply with new federal permitting requirements for air emissions. All regulations to which a facility was subject were required to be listed in these federal permits, and if any of the regulations were amended, compliance with the new/changed requirements was necessary with thirty days of the effective date of the change.

Process industries realized that continuous reconnaissance of applicable state and federal rules would be needed to ensure continuous compliance with their new federal permits. To meet this challenge, a regulatory management of change ("RMOC") subscription service was launched. It included reconnaissance of state and federal citations on agency websites, capabilities to conveniently update subscriber databases, and a quarterly notification to each subscriber summarizing the changes and their potential impacts. The historical text of changed rules was automatically stored in each subscriber's database, an especially useful feature due to lengthy lag times that are common in most agencies' enforcement of environmental rules.

Although efficient procedures and computer systems were implemented to achieve these results, manual labor was involved in downloading citation text from the web and in parsing the rule text into a "Before" and "After" snapshot of each change. Performance of the reconnaissance elements of the RMOC system became an issue after the number of scanned citations exceeded 10,000 rules. In response, the computer system was reengineered to scan and extract data from federal and state websites more efficiently. This eliminated much of the manual labor that was involved and boosted performance back to acceptable levels.

When the number of scanned citations reached 30,000, it became clear that more efficient data processing would be needed to meet future increases. A project was launched to devise a new technology for performing RMOC services at much higher loads. This research led to the development of a new web service, soon to be announced, that will not only meet foreseeable demands but can also fully automate the service.

Of special note in this example is that the new technology is not simply a refinement of the prior system tools. It is a completely different and much faster way to scan information from websites using tools and techniques that did not exist when the original RMOC service was launched. This quantum leap in capabilities enables the new system to handle all of a subscriber's applicable rules, not just environmental ones, and to notify key personnel when changes were detected. Moreover, regulations written in multiple languages may be scanned by the new system to detect possible changes on a daily, or even hourly, basis.

The point of this example is to emphasize that "necessity is the mother of invention." The decision to initiate a service using available technology and system tools, and not wait for better ones before taking action, has led to an RMOC solution that supports even "ultimate" capacity usage of the system with relative ease. Though load-limited and known to be so from the start, the initiating technologies served a purpose critical to the chain of innovation.

Similarly, the implementation of Smart Plant technologies will occur more than once in future cycles of reinvention and reengineering. Pursuing the adoption of Smart Plant alternatives now, even though future cycles may not be fully supported by technology, may be a successful strategy.

Systematic Engineering Drives EPC Project Success

Syamal K. Poddar Ph.D., P.E., Fellow AIChE
Principal Consultant
Poddar & Associates

For a typical project involving Engineering, Procurement and Construction (EPC), the Total Installed Cost (TIC) is split between the above three phases as 10 to 20% , 35 to 45% and about 40 to 50% respectively. Therefore the common notion to manage the project effectively in order to control the budget and thereby to avoid any over run is to focus heavily on the Procurement and Construction phases of the project because P&C are two significantly higher cost components of any EPC project. However, the cost influence of E on an EPC project most often could be very significant. It is therefore important to understand and appreciate that poor and sloppy engineering could potentially spell disaster for any EPC project. Such importance needs to be taken into account at the very beginning of the development of project execution plan.

In order to perform engineering phase in a focused and systematic fashion many operating companies and EPC companies have their own way of performing this phase. However, IPA's Front End Loading (FEL) approach has become an well accepted industry practice. Following the FEL process this paper elucidates the necessary steps for a systematic approach to perform Engineering phase from conceptual to detailed engineering of any given EPC project. Following such a rigorous systematic methodology helps to avoid any adverse impacts on the remaining two significantly higher cost components of the project and thereby to achieve a successful project. The influence of various phases of any EPC project on the project cost is elucidated in Figure 1 below. This plot shows how important it is to do the front end effort right. Once that phase is completed the influencing factor of the subsequent phases of the project to create the opportunity to minimize cost over run as well as project schedule of the overall project gets eliminated.

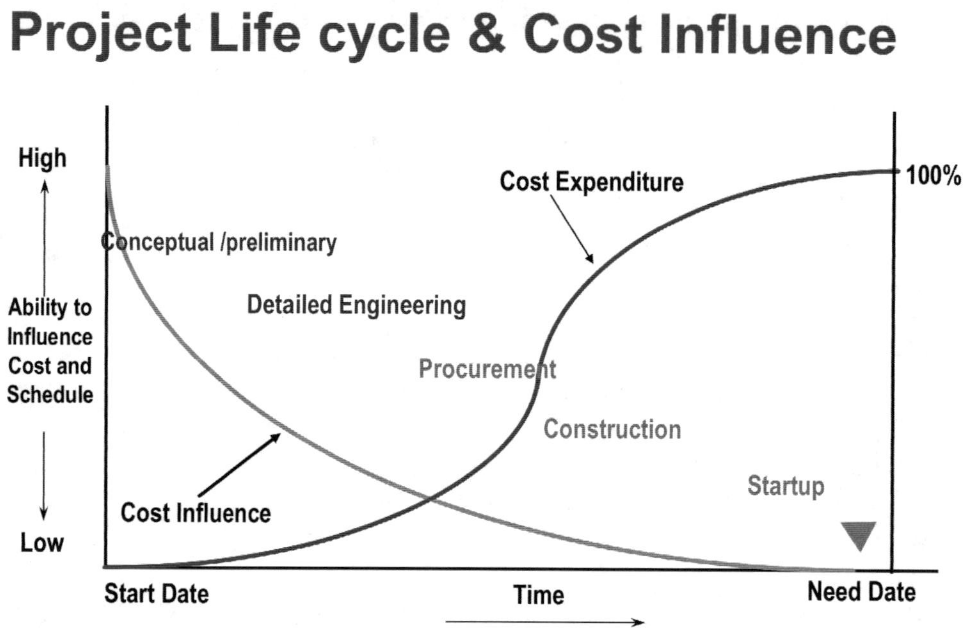

Figure 1

On a given project, FEL covers all the basic and preliminary engineering prior to commencing the detailed engineering. Depending on the complexity of the project, this exercise accounts for 15 - 25 percent of total engineering effort. A proper front end engineering effort sets the stage for an effective detailed engineering

FEL helps to achieve a detail definition of scope of any project. It establishes the business objectives of any given project. It translates business and technology opportunities into a capital project. It aligns all stakeholders to the defined scope and execution plans. It ensures total commitment to a no change philosophy once FEL is accepted by all stake holders.

There are three phases of FEL:

FEL 1
FEL 2
FEL 3

FEL 1 defines the project business case
FEL 2 Phase establishes project/process concepts and
FEL 3 Phase produces detailed engineering and project execution plan

Once executed properly, this sort of engineering effort phase creates the following three categories of deliverables:

1) Engineering documents such as PFDs, P&IDs, plot plans, energy and mass balances, single line electrical diagrams, and specifications of major equipment,
2) Project execution strategy
3) A cost estimate typically at the ±20% range, often as high as ±10%

Typical barriers to do systematic front end planning & engineering:

Depending on the culture of the project owner and the contractor several issues may surface to shorten the project planning and front end engineering efforts. Such detailed effort sometimes is viewed as a waste of time and money. Sometimes the ego and attitude of project manager and/or key management personnel comes into play: "we know how to do it, been there done that (btdt), no room for innovative ideas"

In order to preserve the benefits of proper front end engineering and planning a detailed transition step is to be incorporated in the project execution plan which should include a written transition plan to detailed engineering, procurement and construction phases. this should include division of responsibility (DOR) and detailed project schedule. Utilization of adequate gate checklist to ensure completion of FEL and transition to detail engineering is important. There should be a concerted effort to ensure continuity of key resources in the project team. This is critically more important to maintain the continuity of key process engineers beyond the front end engineering phase. To achieve success in executing an EPC project continuous and consistent interaction & input between engineering and construction representatives are a must.

Concluding remarks:

Proper front end engineering and planning as shown in Figure 2 followed by a properly executed detailed engineering phase integrated with a well thought of transition to procurement and construction phases will result in minimization of engineering rework and thereby a successful project execution.

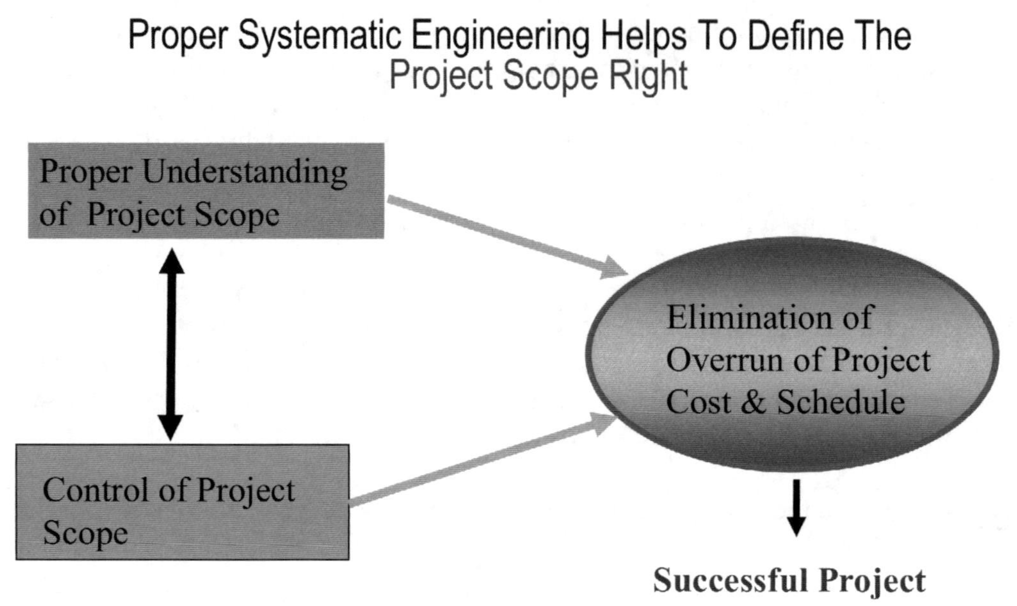
Proper Systematic Engineering Helps To Define The
Project Scope Right
Proper Understanding of Project Scope
Control of Project Scope
Elimination of Overrun of Project Cost & Schedule
Successful Project

Figure 2

Managing Project Risks

Dr. Eldon R. Larsen, Marshall University
Mr. Ryland W. Musick, West Virginia Division of Highways

Abstract

Nearly all projects have risks, both known and unknown. Appropriately managing these risks in a well-thought-out, planned, methodical way can very often make the difference between project success and failure. This paper discusses some common ways to manage project risks, based on good project management concepts.

Introduction

Project managers need to pay constant attention to project-associated risks. Project risks are uncertain events or conditions which could impact that project. These uncertain events or conditions can be categorized as either *known unknowns* or *unknown unknowns*. Some risks we can foresee, while others we cannot. Some risks we can control, while others we cannot. In any case, there will always be project risks.

Known unknowns are the uncertainties about which we know we should be concerned, but we might not know enough about their impact on cost, schedule, and scope, their likelihood of occurrence, how to prevent them, how to make the desirable ones more likely to happen, and so forth. But we know we don't yet know these characteristics, and we can now plan to do something about these unknowns.

Unknown unknowns, by contrast, are events or conditions that we have no realistic way of expecting that they might occur—we don't even know enough realistically to ask any specific questions about them because they are completely unexpected to have an impact on the project and we don't even know what they are—they are unknown unknowns.

Risks can be good or bad, opportunities or threats, positive or negative. These risks can impact project scope, duration, costs, stakeholder relations, quality, and so forth. We want to maximize the likelihood that positive risks will occur. By contrast, of course, we wish to minimize, prevent, or eliminate the potentially negative impact of threats.

Formal project risk management includes planning for risks, identifying risks, analyzing risks, responding to risks, and monitoring and controlling risks.[1] In addition, the attitudes of the performing and requesting organizations toward risks can be a very important factor in risk management. In almost all organizations there must be a balance between a willingness to take risks and a desire to avoid risks. In all cases, excellent communication will be essential to good risk management.

From an insurance standpoint, we sometimes categorize risk into two categories: *pure risk* and *business risk*. Pure risk is the type of risk for which we can buy insurance, and the risk is transferred to the insurance company. Business risk is the type of risk that an organization assumes as part of its ongoing business. For example, the risk that we might not be able to sell all of our products would normally be considered business risk.

Plan for Project Risks

Good risk management planning can dramatically increase risk management effectiveness and efficiency. This is particularly true early in project planning, but applies throughout the project. Planning for risk requires us to be focused and attentive to the project from every angle. When planning for risks, we must remember the good solid principles that have worked in years past, but have the ability to learn new principles that have not previously been applied.[2] The objectives of risk management planning are (1) to decide upfront how to approach, plan, and execute risk-management activities throughout our project, (2) then to update those plans continually as we execute, monitor, and control project risk management activities, and (3) to establish an agreed-upon basis for project risk evaluation.

Of course, all of this risk management planning needs to be balanced against the project's needs, the project's importance, and the needs of the organization. There is no need to kill a tiny-project fly with a sledge hammer, but we do need a sledge hammer to break up the concrete of highly expensive, major project unknowns. If we don't manage the unknowns, they will end up managing us!

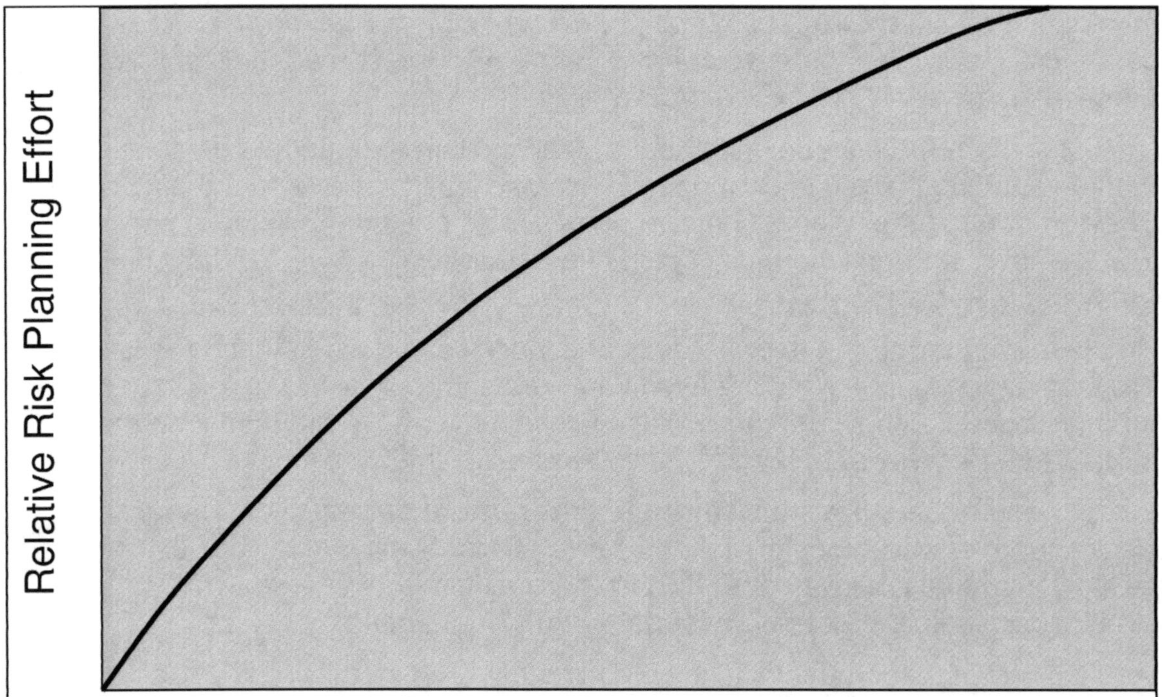

Figure 1: Relative Importance of Project (Project Priority and Cost)

The effort and techniques we apply to the risk management planning will be affected by the culture of the organizations and the attitudes of the people involved on the project. Often there are organizational policies and standard procedures that we are required to follow. If this is the case, it is very important to become thoroughly familiar with those policies and procedures, as well as being careful to develop excellent communication channels with the key risk stakeholders.

When planning how we will management project risks, it is always important to learn all we can from experts who can provide insight, experience, and good judgment, especially with respect to the particular project at hand. We draw on their expertise especially during *risk-planning meetings*. During these meetings we should (1) define our basic risk-management plans, (2) define the risk-management terminology we will use, (3) identify project activities that will deal with risks that could affect project cost, schedule, and scope, and (4) make specific risk management activity assignments to specific people. These planned project activities all become part of the overall project plan, and therefore part of the cost baseline, schedule baseline, and scope baseline. These schedule activities will of course require resources and time to complete, and therefore they affect resource loading and leveling concerns.

One part of risk management planning is to establish project *risk categories*. These categories establish structure for the desired level of detail as we break down the risks into smaller and smaller, better defined risk elements. These risk categories should be reviewed and changed as needed throughout the risk identification and definition process. For example, we might develop risk categories such as technical risks, performance risks, cost risks, schedule risks, management risks, organizational risks, external risks—such as market risks, competition risks, and regulatory risks—and any other risk category that serves this particular project.

Another part of planning for risk management includes the need to clearly define what we mean by various levels of risk likelihood and risk impact. Some example *risk likelihood categories* might include *very low, low, moderate, high,* and *very high*. Alternatively, we might assign numerical weights to those same categories, such as 0.1, 0.3, 0.5, 0.7, and 0.9 respectively. A similar procedure would apply to the specification *risk impact categories*. While doing the risk impact categories, remember that risk impacts can be positive or negative, so both positive and negative numerical weightings might be assigned to those categories accordingly.

When defining the risk likelihood categories and the risk impact categories, it is evident that we need to clearly define what we mean by each category, so that classification of the various risks will be as straight forward and agreed-upon as is realistically possible. An important consideration when developing these likelihood and impact category definitions is to consider defining the categories as functions of the project deliverables.

Table 1: Example definitions of *Risk Impact Categories*

Project Objective	**Impact = 0.8-1.0, or Very High**	**Impact = 0.6-0.8, or High**	**Impact = 0.4-0.6, or Medium**	**Impact = 0.2-0.4, or Low**	**Impact = 0.0-0.2, or Very Low**
Cost	> 50% increase	20-50% increase	10-20% increase	5-10% increase	<5% increase
Time	> 30% increase	20-30% increase	10-20% increase	5-10% increase	<5% increase
Scope	Change is very large; kill or new project	Major change in objectives	Medium change in objectives	Small change in objectives	Very small or no change in objectives

Identify the Project Risks

Project risks can originate from many sources or situations, such as (1) external unpredictable risks—like regulatory or natural forces, (2) external predicable risks—like market risks, operational risks, inflation, or taxes, (3) internal non-technical risks—like management decisions, cash flows, or cost and schedule constraints, (4) technical risks—like changes in technology, design risks, or complexity risks, and (5) legal risks—like licensing risks, patent rights, contractual risks, or *force majeure*.[3]

As we attempt to identify project risks, our objective should be to determine risks that could affect the project, and then to clearly and fully document those risks and their characteristics. This is best done by experts, those who will be doing the work, and other key risk-affected or risk-associated stakeholders. This can be done in a combination of one-on-one meetings with the project leader, or better still as part of a risk planning meeting. The latter gives a chance for everyone to hear the various perspectives and to better achieve proper risk identification and definition. The process can be iterative between risk planning meetings and more focused expert interactions. The result of this iterative process will be the beginning of a *risk register*, which will list the project lists and their characteristics. This will prepare us for the next steps of analyzing risks, responding to risks, and monitoring and controlling risks.

To identify project risks, we use several resources and procedures. (1) We need to learn from past project plans, files, and risk documents. We need to learn from the successes and failures of past projects, as these successes and failures are often associated with how well or how poorly project risks were handled in the past. (2) We can use the Delphi technique of iteratively surveying experts and managers iteratively and anonymously to gain consensus on project risks, likelihoods, and impacts. (3) We can interview team members and other subject-matter experts and those who will be doing the work, as well as other key stakeholders. (4) We can use root-cause analysis and cause-and-effect diagram analysis. (5) We can use SWOT (Strengths, Weaknesses, Opportunities, and Threats) analysis. (6) We can use brainstorming sessions to identify potential risks, and then follow the brainstorming with some type of qualitative analysis of the brainstormed risks.

The result of all of this risk identification will be the beginning of the risk register. This register will be expanded during later steps of risk planning. Shown below is an example beginning of a risk register.

Table 2: Example Risk Register Initial Format

Risk No.	Risk Description	Root Cause	Assumptions	Needed Response	Risk Category
1					
2					
3					
4					
5					

Analyze Risks

There are really two major categories of risk analysis: qualitative risk analysis and quantitative risk analysis. Sometimes these two analysis categories are done separately, and sometimes they are done simultaneously. In a few cases we go directly to quantitative risk analysis and skip qualitative risk analysis. However we go about the process, we need to end up with a clear understanding of the risks and their likelihoods and impacts, preferably on a numerically prioritized scale. Once the initial risk analysis is completed, it is important that we continue to review and update the analyses throughout the duration of the project—keep risk analysis alive!

The objective of qualitative risk analysis is to help us to focus quickly and inexpensively on high-priority risks, giving us a rough understanding of both the potential impact and likelihood of occurrence for each important risk. The analysis should look at risk impact on schedule, costs, scope, quality, risk tolerances, stakeholders, and specific project deliverables.

When qualitatively analyzing project risks, we estimate and record (1) risk probability and (2) risk impact, both positive and negative, on the project deliverables. Next, we repeat the process quantitatively. During these analyses, we use a *probability and impact matrix*, an example of which is found in Table 3, to help us prioritize the risk rating for each identified risk.

Kendrick recommends another graphical approach.[2] In this approach, we plot risks on axes indicating their impact on both the project schedule and budget, and indicate the likelihood of the risk occurring by the size of the circle representing the risk. This is illustrated by Figure 2.

Table 3: Example *Probability and Impact Matrix*

Example Risk Rating Priority Level:
0.60-1.00 = 1st
0.40-0.60 = 2nd
0.25-0.40 = 3rd

		Negative Impact (Threats)					*Positive* Impact (Opportunities)				
Probability		V. L. Neg.	Low Neg.	Mod. Neg.	High Neg.	V. H. Neg.	V. H. Pos.	High Pos.	Mod. Pos.	Low Pos.	V. L. Pos.
		0.1	0.3	0.5	0.7	0.9	0.9	0.7	0.5	0.3	0.1
V. H.	0.9	0.09	0.27	0.45	0.63	0.81	0.81	0.63	0.45	0.27	0.09
High	0.7	0.07	0.21	0.35	0.49	0.63	0.63	0.49	0.35	0.21	0.07
Mod.	0.5	0.05	0.15	0.25	0.35	0.45	0.45	0.35	0.25	0.15	0.05
Low	0.3	0.03	0.09	0.15	0.21	0.27	0.27	0.21	0.15	0.09	0.03
V. L.	0.1	0.01	0.03	0.05	0.07	0.09	0.09	0.07	0.05	0.03	0.01

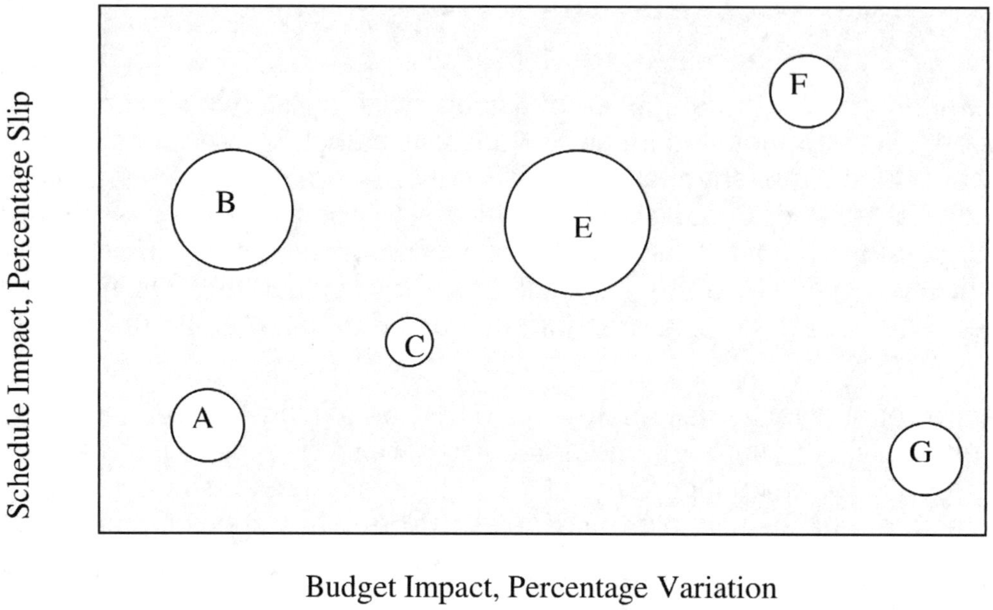

FIGURE 2: Example plot of project risk effects and probabilities

We use the results from the qualitative and quantitative risk analysis to update the risk register with respect to: risk ratings, relative risk priority levels, response time frames, risks needing additional analysis, observations and trends, and even a list of less important risks to monitor, just in case they become important later on during the project. An example is shown in Table 4.

Table 4: Example Columns of Information in an Updated Risk Register, Following Initial Risk Analysis

Risk No.	Risk Description	Root Cause	Assump-tions	Needed Response	Risk Category	Risk Rating	Priority Level	Response Time Frame	"Watch" or What to Monitor?	Need Additional Analysis	Additional Observations & Trends
1											
2											
3											
4											

Quantitative risk analysis can be aided by (1) interviewing of experts, (2) using probability distributions, (3) doing sensitivity analyses, (4) using Monte Carlo or other modeling and mathematical relationships and procedures, and (5) doing expected monetary value calculations.

Plan Risk Responses

The objectives of risk response planning include (1) determining the best options and backup plans to maximize our project opportunities and to minimize our project threats; (2) identifying those who will take responsibility for each important risk, including the possibility of passing some risks to an insurance company; (3) prioritizing risk responses; (4) adding risk response activities into the project plan; and identifying conditions that trigger execution of contingency plans, when conditions indicate that a particular risk is likely to occur in the near future.

When planning responses for negative risks—threats—there are several common approaches: (1) avoid the threat, (2) transfer the risk for the threat, and (3) mitigate the consequences of the threat.

To avoid the threat, we can (1) try to take some action which will eliminate the risk, (2) isolate the project from the threat, or (3) change the project objectives or constraints that are vulnerable to the threat.

To transfer the risk for the threat, the most common action is to purchase insurance for the particular negative consequence, if the threat occurs. This simply transfers risk ownership from us to the insurance company, within the constraints of the insurance contract.

Mitigation of project risk means that we acceptably reduce the impact and/or likelihood that a given threat will actually occur. We want to do this as early as possible—there is a reason for the saying, "An ounce of prevention is worth a pound of cure." For example, we might try to reduce project complexity, do more detailed research before certain decisions are made, choose better, more reliable resources, or develop redundancy in high risk situations.

When planning responses for positive risks—opportunities—there are several common approaches: (1) exploit the opportunity, (2) share the opportunity with someone who has the ability better to help us to achieve the benefits of the opportunity, and (3) grow the likelihood that the opportunity will present itself.

Finally, for both positive and negative risks, one response strategy is to simply allow the threat to occur and plan contingencies accordingly, which will be activated if the threat materializes. Again, risk triggers should be identified during response planning so that we know as quickly as possible that a risk is about to actually happen.

Monitor and Control the Risks

Throughout the entire project we must continually monitor and control project risks. This means that we must also continue to plan and re-plan throughout the entire project. We need to monitor for risk triggers, identify when risks have occurred and whether or not the planned responses were satisfactory, make planning changes for risks that were not satisfactorily handled, and monitor any secondary risks that might occur as a result of a different risk occurring.

All of this means that we must pay constant attention to the project objectives, stakeholder expectations, project baseline deviation trends, contingency reserves, and new risks that might be developing or less important risks that might be changing in priority. Therefore, project risk needs to be an agenda for most, if not all, of the project meetings. We should also be monitoring how we as a project team are doing at following and executing risk response plans and procedures. When necessary, workaround plans should be developed—these are plans nor originally planned, but developed as a result of new risk developments that were unexpected. These workaround plans should be carefully documented, to maintain the integrity of the risk management procedures and so as to allow lessons learned to be developed.

The risk register needs to be regularly updated as a result of our monitoring and controlling risks and risk responses. The changes to the risk register should be carefully documented, and we should carefully note any lessons learned for future use during our project or future projects.

Conclusion

Risk management is an essential part of project management! The tools, principles, and concepts presented in this paper are only an overview of the many important aspects of project risk management, but they are essential to good project risk management. To neglect good risk management is to endanger the very success of our projects.

References

1. *A Guide to the Project Management Body of Knowledge*, Third Edition, Project Management Institute, Inc., 2004.
2. Kendrick, Tom. *Identifying and Managing Project Risk: Essential Tools for Failure-Proofing Your Project.* AMACOM, 2003.
3. Wideman, R. Max. *Project and Program Risk Management: A guide to Managing Project Risks and Opportunities.* Project Management Institute, Inc., 1992.

Authors

ELDON R. LARSEN is a professor of engineering at Marshall University Graduate College (100 Angus E. Peyton Dr., South Charleston, WV 25303; Phone: 304-746-2047; Email: Larsene@marshall.edu) where he coordinates the M.S. in Engineering degree program. He has taught project management courses since 1994 in industrial and university settings; he also teaches operations management and management of technical human resources and organizations. Before joining Marshall full-time in 1999, he worked for 16.5 years at Union Carbide Corporation, where he led a major effort to develop and implement project management for research and development. He earned B.S. and M.S. degrees from Brigham Young University, and a Ph.D. from the University of California at Berkeley, all in chemical engineering. He is a senior member of AIChE, where he has served twice as chair of the Charleston local section. He has also served as chair of the AIChE Management Division, and served as chair of the Planning and Organizing committee for the 2007 AIChE/ACS Management Conference, and served as vice chair of the conference. He is a past President and CEO of the West Virginia/Ohio Valley Chapter of the Project Management Institute (PMI).

RYLAND W. MUSICK, JR., PE., is a design engineer for the West Virginia Division of Highways in Princeton, WV. He earned a B.S. in civil engineering from the West Virginia Institute of Technology in Montgomery, WV, and an M.S. in engineering with an emphasis in engineering management from Marshall University in 2009. He is currently pursuing a Ph.D. in transportation infrastructure at Virginia Polytechnic Institute and State University in Blacksburg, VA.

Title: Effect of an improvisational approach to new product development during a crisis: an empirical study of NPD teams in the field of chemistry and chemical engineering.

Authors: Yassir M. Samra, PhD
Jonathan B. Hartman, PhD
Gary S. Lynn, PhD
Richard R. Reilly, PhD

Abstract: For many years, past studies have supported the hypothesis that a traditional new product development process increases the odds of faster product launch and greater new product success. More recent studies have shown that an improvisational approach is also associated with a faster and/or successful new product launch. The two schools of the thought have addressed different environments in which new product development operates in. However, there seems to be a lacuna in the scholarship on new product development when studying the effects of a NPD process as the organization experiences a crisis. The purpose of this paper is to examine the possible moderating effect a crisis can have on the relationship between an improvisational approach and both the speed and the success of the new product launched. Using a specially designed questionnaire, 301 NPD project managers working on products associated with the field of chemistry and chemical engineering were asked their opinion of their team's use of an improvisational NPD process on their most recently completed NPD project, the degree to which they felt the NPD team was operating during crisis, as well as how fast they were in launching the new product and how successful the new product was in the market. The results indicate that while a crisis is positively associated with improvisation, speed and success, it does not moderate the relationship between an improvisational approach and neither speed nor success.

INTRODUCTION

Crises can have a debilitating effect on an organization (as the Bhopal tragedy had on Union Carbide) and yet they also can have an energizing positive effect on a firm (as the Tylenol scare had on Johnson & Johnson). Several studies have explored different events that have threatened the existence of an organization to find characteristics of both successfully and unsuccessfully resolved crises. As crises are threatening in nature, successful organizations react to a crisis quickly to abate the situation while unsuccessful organizations do not react fast enough to control the situation or do nothing at all hoping the situation will eventually resolve itself.

A crisis is a situation that was never predicted nor recognized as a threat until the event actually occurs. At this point, the employees of an affected organization perceive a high threat level to the stability of the organization. Such a threat can have a unifying effect on the employees of an organization to fight harder to save the organization from peril. In most cases, the crisis will financially subdue the organization (as Union Carbide eventually was bought out and Johnson & Johnson spent over $200 million in a national recall). Crises generally fall into three categories: disasters (Bhopal tragedy), scandals (downfall of Enron), and product safety (deaths attributed to Tylenol). In addition to these three categories, an organization may experience a crisis as its position becomes threatened from the onset of unanticipated competition (like at Iomega) or from innovative products that do not meet the criteria of the market (the mathematical flaw in the Pentium chip). In this situation, new product development can be the answer to a crisis as it can generate a new profit stream to prolong the life of the organization and deliver it from the brink of extinction. The employees placed on a new product development team, as members of the organization, can have a pivotal role to play in delivering the organization from the crisis.

For many businesses, NPD is essential to the growth (Cooper, 1984) as companies need to generate new products and services (Lynn & Reilly, 2002) to create or sustain a competitive advantage (Brown & Eisenhardt 1997; Mascitelli 2000). The NPD literature offers both a traditional (phased-review) and improvisational processes and empirical studies in NPD have explored outcomes of new projects under conditions of turbulence and uncertainty but surprisingly, the NPD scholarship has just begun to explore the phenomena of crisis and empirically assess its impact on the success of NPD projects (Akgun et al., 2006, Samra, 2005). Since a crisis is something that cannot be planned for, an improvisational approach to NPD seems more appropriate than a traditional one.

While the scholarship on crisis has viewed its impact on the organization as a whole, this study seeks to extend Tjosvold's (1984) work on the perception of a crisis at the team level by using NPD teams and their perceptions of an organizational crisis. In addition this study will extend Eisenhardt & Tabrizi's (1995) work on the use of different approaches in turbulent and uncertain environmental conditions by investigating the impact of an improvisational approach in a crisis situation. Using a data set of respondents across different industries in the chemical sector, an empirical test will determine whether an improvisational approach during a crisis can lead to better project outcomes.

LITERATURE REVIEW

Crisis

Most crises occur through a triggering event. Some scholars claim that crises are disastrous situations that severely threaten an organization. Specifically, Starbuck et al (1978) conjecture that crises originate as threatening events from either an organization's external environment or from defects within the organization itself. King (2002) claims that a crisis is an "unplanned event that has the potential of dismantling an organization" (p. 237). Pearson & Mitroff (1993) state that a crisis is a situation when the survival of the whole organization is in jeopardy. Also Pearson & Clair (1998) define an organizational crisis as a highly ambiguous infrequently occurring situation where causes and effects are unknown but threaten survival.

Another component of crisis is the sense of urgency to respond so that the situation does not continue to threaten the organization. If an organization is constantly facing adverse situations then it will grow accustomed to this environment. Yet any organization can experience a crisis (Shaluf et al, 2003). In fact a crisis is not simply an adverse situation but can be viewed as a disruption in time that requires the immediate attention of those in charge (Weick, 1977). Pearson & Clair (1998) assert that a crisis will require little time (if any) to respond to the event and will present the organization with a dilemma to make a decision on how to settle the situation. Others have compared organizational crisis to the event of a fire, claiming that in times of fire, organizations must react quickly to control and subdue the force of the fire before it claims the life of the organization (Smart & Vertinsky, 1984; Weick, 1993; Carter, 1997).

Thus a crisis consists of three parts: (1) a triggering event that (2) surprises and threatens the organization's existence and requires (3) an urgent response in order to change the potentially negative event to a more positive outcome (Hermann, 1963). Billings et al (1980) complement Hermann's model by introducing crisis as a measure of perception that values are seriously threatened, the individuals involved are uncertain that a practical response will work, and a quick response is necessary. But the idea of perceiving a crisis is extremely important as two individuals can perceive the same situation differently. For example, a crisis may be perceived by new residents living in the midst of a forest fire, but is not perceived as such by firefighters as the situation is certainly one in which they have received appropriate training for dealing with. While the former "never expects" or does not anticipate a fire, the latter knows exactly what to do.

Since crises are more subjective in nature, they can be artificially generated to evoke perceptions of a threatening situation. In a study by Kim (1998), the government of South Korea imposed a crisis on the "Korean" car industry. The new policy was to have Korean car companies develop their own models. By initiating a crisis situation rather than react to a triggering event, the Hyundai Motor Company was able to design and launch a Korean car, thereby moving Korean car manufacturers from the typical business of assembling foreign cars to designing their own. While Kim (1998) views crisis as both a danger and opportunity, this does not take into consideration Hermann's (1963) model of loss of values and threat to survival. The Korean car companies apparently did not view their companies as being in a crisis situation according to these criteria as their business of assembling foreign cars would not threaten their survival; but the employees' perception of crisis was high nonetheless. The events do account for a rapid response, in the sense that Hyundai did develop cars rapidly in response to the government imposed crisis. This is also seen in a study by Barnett & Pratt (2000) where the authors found that organizations initiating a crisis on their employees had more successful outcomes. While the distinction can be made between a crisis that poses a latent threat and one that poses a "real" threat, the authors assert that there is no significant difference on the impact of each threat on the employees. In both cases, whether the threat is real or latent, the perception by the employees should be that a threatening situation has surfaced which requires immediate attention.

New Product Development Practices

Common formal processes in NPD tend to be associated with stable environments and with incrementally innovative products. In such times, NPD is predictable and the technology associated with the new product is known. Therefore, a structured approach is appropriate. But NPD isn't always straightforward. Scholars have suggested that sequential models may be too general to fit the demands of some products and services. For instance, structured models may be inappropriate for products requiring extraordinary speed, secrecy, address specific problems, or entail short production runs (Gwynne, 1997). They have also been shown to be rigid, and as a consequence, may reduce flexibility (Rosenthal, 1992). Indeed, scholars have suggested that they may be too structured for quickly changing competitive environments (Cooper 1994; Hoopes & Postrel 1999). To counter this, two models have been developed. The first is Cooper's (1994) flexible approach to NPD which revolves around being fluid and adaptable; incorporating "fuzzy gates" that are both situational and conditional was empirically tested and supported (Cooper & Kleinschmidt, 1995). The second is an improvisational model (Eisenhardt & Tabrizi, 1995; Moorman & Miner, 1998a); one that allows the NPD team to think in the midst of execution. In a widely cited study on new product development activities at two midsize firms, Moorman and Miner (1998a) found that improvisation can be an effective tool when an organization faces environmental turbulence that requires action.

Improvisation

To define improvisation, a literature review was conducted and various definitions were found. For example, Barrett (1998) describes improvisation as coming up with novel responses without a set plan and Bastien and Hostager (1988) define improvisation as inventing and executing new ideas. This stream states that improvisation is a deviation from normal routines or behavior. This is insufficient in part due to the fact that this is synonymous with the definition of creativity (Amabile, 1996). An example of this is when an organization designs a creative marketing strategy (Moorman & Miner, 1998b). The idea may be new and innovative but it is not necessarily improvisational.

On the other hand, improvisation is synonymous with creativity only under a time constraint or pressured situation. The dictionary defines improvisation from the Latin word "improvisus" meaning "unforeseen" as composing extemporaneously. Thus a new set of definitions for improvisation is found. For example, Crossan & Sorrenti (1997) define improvisation as intuition guiding action in a spontaneous way. Cunha et al (1999) define improvisation as the conception of action as it unfolds. Perry (1991) defines improvisation as formulating and implementing strategies in real time which is echoed by Weick (1993) as improvisation has no distinction in time between composition and performance. Moorman & Miners' (1998b) definition of improvisation states that improvisation is when the planning and execution converge in time so that they occur simultaneously. In the example provided by Moorman & Miner (1998b) of a CEO responding quickly to a crisis, there is still a time gap between planning the response and executing the response to the crisis. Therefore, planning and execution are two discrete events (that do not necessarily converge in time). But the response is asserted to be improvisational as the amount of time that the CEO has is limited.

The jazz literature asserts that improvisation is composing on the spur of the moment. However, musicians may compose a solo that does not significantly differ from what has been played before; thus lacking creativity. Therefore, when a musician plays a jazz solo that (s)he has performed before; it meets the second criterion for improvisation but fails to meet the first. At this point, it is safe to say that an acceptable definition for improvisation contains both factors (creativity & spontaneity) and this is congruent with Zack's (2000) definition which states that improvisation involves maximal innovation in a short period of time; thereby combining both the creativity as well as simultaneity. This definition is in agreement with a later definition provided by Miner et al (2001) in which the authors still define improvisation as the degree to which composition and execution but accept the notion that pre-existing routines do not account for improvisation.

HYPOTHESES

Crisis & Improvisation

In threatening, situations, it is all too easy to rely on what one knows or what one has been trained to do (Barthol & Ku, 1959). This is seen in the airline industry where a captain does not rely on individual action but rather on the contingency plans that have been developed for the specific circumstance. Evaluation of pilot error accidents have found that the situation was (in some cases) worsened by relying on individual action (Disaster Database, 2002). Heath (1995) proposes that the more an organization experiences disasters, the more routine will be their response. The situation of an airline's engine failure indeed threatens the lives of those on board, but prior planning for this circumstance (and others) can limit the damage (Quarantelli, 1988). Yet it is impossible to have a structured response for every contingency since the number of possible crises is virtually infinite (Weick, 1988). In this dangerously unfamiliar situation, some degree of trial and error is present and as Bateson asserts: "An explorer can never know what he is exploring until it has been explored" (Weick, 1988: 305). Therefore, normal reactions to a crisis do not necessarily work because of the entirely new situation the organization is facing. In fact, Starbuck et al (1978) claim that a situation cannot be deemed a crisis if normal behaviors produce improvements. Therefore, in returning to the airline example of engine failure, it is realized that this situation should not be a crisis for the airline crew but perhaps is perceived as a crisis to its passengers.

Another example of a crisis event was the infamous Mann Gulch fire (Weick, 1993) where many of the firefighters perished. At first, the team of firefighters attempted to pass the gulch towards a river that would lead them to safety. As the flames quickly approached, the team leader, Dodge, decided to change direction and attempt to lead his crew up a steep hill to avoid the approaching flames but was unsuccessful. After relying on logical solutions, the final act (of desperation) of Dodge (and to the amazement of his crew) was that he ordered his crew to abandon their firefighting tools and lit a fire in front of them and ordered them to lie in this ring of fire with him. No one heeded their superior's call and while they tried to outrun the fire, only two other members had survived unburned (a third survived but due to his burns, died the next day). It took 450 men and five days to get the 4,500 acre Mann Gulch fire under control, a fire that was originally classified as being between 10 to 99 acres. The Forestry Service held an inquiry and determined that had the crew obeyed Dodge's order to lie in the escape fire, they would have been saved.

The reason for a detailed description of this event is to illustrate that high-pressure environments are generally considered to be infertile grounds for improvisation (Cunha & Cunha, 2001) as they do not provide sufficient time to think creatively (Amabile, 1996). Weick (1993) asserts that when people are put under pressure, they respond in their most habituated ways. He continues by saying, "What we do not expect under life-threatening pressure is creativity" (Weick, 1993: p. 639). Therefore we can conclude that under times of crisis, the one thing we don't expect is improvisation, however if it is done, it can be very rewarding.

The NPD literature has yet to empirically establish an association between crisis and improvisation. While several factors associated with new product success have been observed in turbulent environments, they have not been tested in crisis situations. It is important to understand the distinction between turbulence and crisis. The dictionary defines turbulence as a state of unrest or disturbance, while crisis is defined as an unstable or crucial time or state of affairs in which a decisive change is impending, thereby implying that a crisis has a much higher degree of threat. Industries that are constantly turbulent and the successful firms in the industry have acclimated themselves to change (Brown & Eisenhardt, 1997) and therefore turbulence becomes part of their doing business. Crisis on the other hand can develop from a steady state. When a crisis does occur, the rules essentially get thrown out the window (Weick, 1993) and the NPD team has to arrive at novel solutions quickly. Thus as the literature accepts that improvisation can be a useful tool throughout NPD and a crisis, the following is hypothesis is presented:

> *Hypothesis 1: For a firm in crisis, higher threat levels as perceived by the NPD team will be positively associated with higher levels of improvisation by the NPD team.*

Crisis & Outcomes

Since a crisis is a threat to survival, urgent (re)action is necessary to abate the situation. An organization typically has neither the luxury nor the time to analyze several responses nor to develop a manner with which to execute them in hopes of delivering itself from the crisis; rather the crisis will worsen the situation if action is not swiftly taken.

In sports (particularly American football) there are countless moments when a team sits on the brink of elimination with little time left to score and manages to overcome enumerable odds to become victorious in the most unorthodox fashion (Katz, 2001). One may argue that in sports, the perception of a crisis is not as threatening as it might be in an organizational setting (lives aren't being lost and there's always next season). But the lesson is still the same; in a crisis situation, time is limited before the situation exacerbates and all is lost. A perceived crisis can be successfully resolved by immediately addressing the threatening nature in hopes of a successful resolution. A company can prepare only so much for a crisis but it can never eliminate the possibility of one occurring. Therefore, if a crisis does exist, then a rapid response is required to quell the situation. If left unattended, the crisis can have further detrimental effects on the organization.

In NPD, crisis can be a good thing as it presents the NPD team with an opportunity to shine and deliver the organization from its (financial) peril. To do so, the NPD team must react immediately to the crisis and develop (and launch) this new product quickly. Formally stated:

Hypothesis 2a: For a firm in crisis, higher threat levels as perceived by the NPD team will be positively associated with speed to market..

In addition, many NPD studies have shown a strong correlation between speed and success. Therefore, if indeed a crisis is a good thing, then it will be positively correlated with both speed and success. Thus the following is hypothesized:

Hypothesis 2b: For a firm in crisis, higher threat levels as perceived by the NPD team will be positively associated with higher levels of success in new product development and launch.

Improvisation & Outcomes

With regard to the NPD scholarship, it has been empirically tested that proficiently using a structured approach will yield positive outcomes (Cooper & Kleinschmidt, 1986, Cooper & Kleinschmidt, 1987; Lynn et al, 1999; Millson & Wilemon, 2002; Samra et al, 2008, Shepherd & Ahmed, 2000). However, based on the preceding arguments, if this is the standard traditional approach to NPD, then it should not have any positive significance on the outcome of a new product in a crisis situation. On the other hand, there is support for the use of improvisation throughout new product development in uncertain and turbulent environments (Brown & Eisenhardt, 1997; Eisenhardt & Tabrizi, 1995). But a crisis is far more threatening in nature than a turbulent or uncertain environment because it contains not only an uncertainty factor but also the perception that the entire organization's survival is in question from this perilous situation. In fact, turbulent environments can be a part of the NPD team's industry (Eisenhardt & Tabrizi, 1995) as the more exposed the NPD team is to change, the more likely they will be able to adapt to rapid changes associated with turbulent environments.

As for uncertain environments, they are the midpoint between a turbulent environment (where changes can be anticipated) and crisis (where the threat level is extremely high and imminent). Uncertain environments are uncharted for NPD teams and differ from turbulent environments as they are not simply changes that the team can adapt to rather they are environments where the NPD team must decide if they wish to enter. In a crisis, the company in question is in peril and it is imperative to respond to the situation at hand. The NPD team can still function in this newly created uncertain environment, but like the rest of the organization, it may feel threatened by the onset of the crisis. As successfully resolved crises require creative answers, the NPD team can improvise to be successful. As previously mentioned in the preceding hypothesis, both speed and success are highly correlated and since a crisis can occur at random to any organization and it is completely unpredictable and unrecognizable as a potential threat, the following hypotheses are presented:

Hypothesis 3a: A firm in high crisis will exhibit a stronger relationship between improvisation and speed in the NPD process than a firm in low/no crisis.

Hypothesis 3b: A firm in high crisis will exhibit a stronger relationship between improvisation and new product success than a firm in low/no crisis.

METHODOLOGY

Hypotheses 1 & 2 suggest that as the perception level of a crisis increases, so does the frequency of improvisation as well as the likelihood of both a faster and more successful product launch. To test this, a bivariate correlation matrix will indicate any support for these hypotheses. Hypotheses 3a & 3b will require using a hierarchical regression model with three blocks. The first two will contain the variables improvisation and crisis, respectively, while the third will contain a new variable that consists of the cross-product between improvisation and crisis. If a significant result is found in the third block, then a new dichotomous variable will be created to differentiate between a firm in high crisis and one in low crisis. Finally a new set of regressions should be able to empirically demonstrate that improvisation is more closely related to positive outcomes during a crisis than not.

Sample

To test our hypotheses, a questionnaire was developed based on previous research (Billings et al, 1980; Cooper & Kleinschmidt, 1987; Cooper, 2001; Davenport & Prusak, 1998; Day 1994; Locke et al, 1981; Lynn, 1998; Moorman & Miner, 1998a & b; Schein, 1993). After designing and refining the questionnaire, we selected a contact person in a variety of US based chemically-related companies (based on the North American Industry Classification System of the US Census Bureau) to participate in this study. We asked the contact person in each company to select a project manager or senior team member (respondents are primarily product/project managers, senior team members or department managers and directors) who was with the project from pre-prototype through launch. Lukas and Ferrell (2000) and Podsakoff and Organ (1986) found that managers rely on their own self-reports and provide reliable and objective data. Also, Huber and Power (1985) note that simply averaging multi sources is less likely to be accurate than using a key informant. After the selection of the respondents, they were informed that their responses would remain anonymous and their responses will not be linked to a company or product name. This increased the motivation of informants to cooperate without fear of reprisals. To improve the accuracy of retrospective reports, recent projects were selected to eliminate the elapsed time between the events of interest and the collection of data. Of the 301 "contact people" asked to participate, 244 of them returned a questionnaire (an 81.1% response rate), of which 55 had sufficient data to be included in both the correlation and regression analyses.

Measures

For this study, questions were measured on a Likert-type scale from 0 = strongly disagree to 10 = strongly agree. Table 1 provides a summary of the measures.

Table 1: Summary of Measures

Dependent Variable	
Success	To operationalize new product success, six questions were asked referring to how well the project met volume, sales, profit, ROI, and market share expectations. All items loaded onto one factor and the mean was used as the variable. References: Cooper, 1994; Cooper, 2001
Speed	To operationalize speed, four questions were asked. Since a multi-company and multi-industry sample was used, the speed-to-market differences in the nature of projects were controlled by using relative measures. This approach and item content were similar to that of Kesslar and Chakrabarti (1999). Speed-to-market was assessed by comparing actual performance to pre-set schedules, company standards and similar competitive projects. Reference: Kessler & Chakrabarti, 1999
Independent Variables	
Improvisation	To operationalize improvisation, the following three questions were asked: (1) the team figured out the NPD process as it went along versus following a rigid well-defined plan, (2) the team improvised in developing the product versus strictly following the plan, and (3) the team improvised in commercializing this product versus strictly following the plan. Items were adapted from Moorman and Miner (1998b). The mean of these items was used as the variable. Reference: Moorman & Miner, 1998a & b
Moderating Factor	
Crisis	Three questions were asked to measure the perception level of crisis. Specifically, respondents were asked the level of crisis in within their organization, environment, and with their customers that their current project would help solve. Reference: Schein, 1993

Measure Validity & Reliability

Before testing the proposed model, the structure and reliability of the constructs were assessed. To measure each construct, a factor analysis was performed (on the items asked in the questionnaire) along with a varimax rotation method and Kaiser normalization to validate the number of constructs used as well as to confirm whether the items mentioned for each construct were indeed capturing their respective construct (Tabachnick & Fidell, 2001). After the extraction of components with an eigenvalue > 1.00, four variables remained explaining 78.6% of the variance. A loading factor value of 0.5 and above in each component in the varimax rotated component matrix validated the items as acceptable measurements of the constructs.

Each variable in the model consisted of the average of the items in each component. To ensure the practicality of the measures, an Alpha reliability test was performed. All Cronbach's alphas (see Table 2) were above the minimum acceptable level of 0.7 as recommended by Nunally (1978).

Table 2: Measures and Reliability

Predictor Variable	No. Items	Mean	Standard Deviation	Alpha
Improvisation	3	5.56	2.21	0.76
Moderating Variable	**No. Items**	**Mean**	**Standard Deviation**	**Alpha**
Crisis	3	4.58	2.32	0.74
Dependent Variables	**No. Items**	**Mean**	**Standard Deviation**	**Alpha**
New Product Success	6	5.36	3.02	0.97
Speed To Market	4	5.44	2.64	0.85

RESULTS

To determine if a crisis can present an opportunity, positive bivariate correlations between *crisis* and each of the two outcomes must be found. The results presented in Table 3 illustrate that indeed there is a positive correlation between *crisis* and *improvisation*, thus supporting H1. The results also show that *crisis* significantly correlates with both *speed* and *success*, thus supporting both H2a & H2b. Also it is worth mentioning that s*peed* was significantly related to s*uccess*, which is consistent with past studies (Cooper & Kleinschmidt, 1994 & 1986).

Table 3: Bivariate Correlations (N = 55)

	New Product Success	Speed to Market	Crisis	Improvisation
New Product Success	1			
Speed to Market	0.568**	1		
Crisis	0.197*	0.360**	1	
Improvisation	0.011	0.041	0.294**	1

* p < 0.05; ** p < 0.01

Hypotheses 3a & 3b sought to assess the degree to which predictor variable *improvisation* along with *crisis* explained the variance of project outcomes *speed* and *success*. To find support for the final two hypotheses, two regression models were used to assess the fit of the model and the impact of the predictor variable on both *speed* and *success*. In each case, the first variable in the model included the predictor variable *improvisation*. The next variable in the model was the moderating variable *crisis*. The final variable in the model included a cross-product between *improvisation* and *crisis* to determine any interaction effect between the two variables. If so, the results would indicate that improvisation may be better in times of crisis than in times of stability.

Speed to Market

The variable *improvisation* was found to have a non-significant impact on *speed*. The squared correlation ($R^2 = 0.000$, $p < 0.979$) was not significantly different from zero. On the other hand, the variable *crisis* was found to have a significant impact on *speed* as the squared correlation ($R^2 = 0.157$, $p < 0.005$) was significantly different from zero. The final block in the regression results indicates that there is no significant interaction between *improvisation* and *crisis* ($R^2 = 0.161$, $\Delta R^2 = 0.004$, $p < 0.625$), thus demonstrating no support for H3a.

New Product Success

The variable *improvisation* was found to have a non-significant impact on *success*. The squared correlation ($R^2 = 0.001$, $p < 0.833$) was not significantly different from zero. The variable *crisis* was found to have a marginally significant impact on *success*. The squared correlation ($R^2 = 0.046$, $p < 0.063$) for *crisis* was partially significantly different from zero. The final block in the regression results indicates that there is no significant interaction between *improvisation* and *crisis* ($R^2 = 0.05$, $\Delta R^2 = 0.004$, $p < 0.604$), thus demonstrating no support for H3b.

DISCUSSION & FUTURE RESEARCH

Through empirical examination of outcomes of NPD (speed & success), its proposed antecedent (improvisation), and moderator (crisis), this study demonstrates associations between these variables. Generally, the data support the propositions of this research. Significant relationships were confirmed between crisis and improvisation, as well as crisis and outcomes (both speed and success); however improvisation was not associated with positive outcomes of NPD. These findings increasingly confirm the theory that a crisis is not only something to avoid but can serve as an opportunity to shine and prosper.

The findings suggest openings for managers of NPD teams to promote beneficial outcomes of NPD. As with any effort to launch a product, active encouragement and involvement throughout the difficult time associated with a crisis is important. In times of crisis, levels of anxiety become elevated and NPD team members may hesitate and second-guess themselves. Therefore, management can be involved with the project to ensure that it is on track and tasks are executed proficiently. The fact that a crisis is positively associated with outcomes suggests that there are other factors that can influence the results of NPD.

The results also illustrate that while a crisis may lead to improvisation, this does not necessarily mean that team improvisation will have more of a significant impact on new product success under crisis conditions than not. This finding is surprising as crisis is defined as containing an element of surprise in the sense that it is not anticipated, yet it is extremely threatening and requires an immediate response. On the other hand, the investigated sector for this study was limited to chemical based companies. The industries included petrochemicals, pharmaceuticals, and semiconductors – where due to potentially hazardous chemical exposure and strict quality control standards – improvisation may not be an option in developing a new product. Therefore, it can only be inferred that a crisis may lead the NPD team to improvise, however, the team performed other tasks associated with NPD that would lead to successful outcomes during a crisis.

The primary limitation of this study is that it considered the chemical sector; a sector that is more closely related to goods as opposed to services. The use of an improvisational approach may be more appropriate for new service associated products as reliability does not significantly affect the user's interface with the product. For future research, it is suggested to compare the chemical sector to a more service-based sector such as telecommunications or healthcare. If dissimilar results are found, then perhaps the use of improvisation throughout NPD should be limited to companies that provide services and not goods.

REFERENCES

Akgun, A.E., Lynn, G. S., and Byrne, J.C., 2006. Antecedents and Consequences of Unlearning in New Product Development Teams, *The Journal of Product Innovation Management,* **23(1)**: 73-88.

Amabile, T. M., 1996. *Creativity in Context*, Westview Press, Inc., Chap. 8, p. 232.

Anonymous. 2002. Disaster Database, *Disaster Prevention and Management.* Bradford: **11(4)**: 323-358.

Barnett, C. K. and Pratt, M. G., 2000. From Threat-Rigidity To Flexibility: Toward a Learning Model Of Autogenic Crisis In Organizations, *Journal of Organizational Change Management* **13(1)**: 74-88.

Barrett, F., 1998. Creativity and Improvisation In Jazz and Organizations: Implications For Organizational Learning. *Organization Science* **9(5)**: 605-622.

Barthol, R.P. and Ku, N.D., 1959. Regression Under Stress To First Learned Behavior, *Journal of Abnormal and Social Psychology* **59**: 134-136.

Bastien, D.T. and Hostager, T.J., 1988. Jazz as a Process of Organizational Innovation. *Communication Research* **15(5)**: 582-602.

Billings, R.S., Milburn, T.W. and Schaalman, M. L., 1980. A Model of Crisis Perception: A Theoretical and Empirical Analysis, *Administrative Science Quarterly* **25(2)**: 300-316.

Brown, S. L., and Eisenhardt, K. M. 1997. The Art of Continuous Change: Linking Complexity Theory and Time-paced Evolution in Relentlessly Shifting Organizations. *Administrative Science Quarterly*, **42(1)**: 1-34.

Carter, T., 1997. Crisis Management for Sales Force Managers, *Journal of Professional Services Marketing* **15(2)**: 87-103.

Cooper, R. G., 2001. *Winning At New Products, 3rd edition.* Perseus Publishing, Cambridge, Chap. 2, 10, pp 23-24, 250.

Cooper, R.G., 1994. Third-Generation New Product Processes, *Journal of Product Innovation Management*, **11(1)**: 3 - 14.

Cooper, R.G., 1984. How New Product Strategies Impact on Performance, *The Journal of Product Innovation Management*, **1(1)**: 5-19.

Cooper, R. G. and Kleinschmidt, E. J., 1986. An Investigation Into the New Product Process: Steps, Deficiencies, and Impact, *Journal of Product and Innovation Management* **3(2)**: 71-85.

Cooper, R. G. and Kleinschmidt, E. J., 1987. New Products: What Separates Winners From Losers, *Journal of Product and Innovation Management* **4**: 169-184.

Cooper, R. G. & Kleinschmidt, E. J. 1994. Determinants of Timeliness In Product Development. *Journal of Product Innovation Management*, **11(5)**,381 -396.

Cooper, R. G. and Kleinschmidt, E. J., 1995. Benchmarking the firm's critical success factors in new product development**,** *Journal of Product Innovation Management,* **12(5):** 374-391.

Crossan, M. M. and Sorrenti, M., 1997. Making Sense of Improvisation, *Advances in Strategic Management,* **14**: 155-180.

Cunha, J.V. and Cunha, M. P., 2001. Brave New (Paradoxical World: Structure and Improvisation In Virtual Teams, *Strategic Change*, **10(6)**: 337-347.

Cunha, M.P., da Cunha, J.V., and Kamoche, K., 1999. Organizational improvisation: What, When, How, and Why, *International Journal of Management Reviews*, **1**: 299-341.

Davenport, T. H. and Prusak, L., *Working Knowledge: How Organizations Manage What They Know.* Harvard Business School Press, Boston, 1998.

Day, G. S., 1994. The Capabilities of Market-Driven Organizations, *Journal of Marketing* **58**: 37-52.

Eisenhardt, K. M., and Tabrizi, B. N., 1995. Accelerating Adaptive Processes: Product Innovation in the Global Computer Industry. *Administrative Science Quarterly*, **84**: 84-110.

Gwynne, P. 1997, "Skunkworks, 1990s style", *Research Technology Management*, **40(4)**: 18 - 23.

Heath, R., 1995. The Kobe Earthquake: Some Realities Of Strategic Management Of Crises and Disasters, *Disaster Prevention and Management,* **4(5)**: 11-21.

Hermann, C. F., 1963. Some Consequences of Crisis Which Limit the Viability Of Organizations, *Administrative Science Quarterly*, **8**: 61-82.

Hoopes, D.G. and Postrel, S., 1999. Shared Knowledge, "glitches", and product development performance, *Strategic Management Journal*, **20**: 837 – 865.

Huber, G. P., and Power, D. J., 1985. Research Notes and Communications Retrospective Reports Of Strategic-level Managers: Guidelines For Increasing Their Accuracy, *Strategic Management Journal,* **6**: 171-180.

Katz, N., 2001. Sports Teams as a Model for Workplace Teams: Lessons and Liabilities, *Academy of Management Executive*, **15(3)**: 56-69.

Kessler, E.H. and Chakrabarti A.K., 1999. Speeding up the pace of new product development, *Journal of Product Innovation Management*, **16(3)**: 231-247.

Kim, L., 1998. Crisis Construction and Organizational Learning: Capability Building In Catching-up At Hyundai Motor, *Organization Science*, **9(4)**: 506-521.

King III, G., 2002. Crisis Management & Team Effectiveness: A Closer Examination, *Jounal of Business Ethics*, **41(3)**: 235-249.

Locke, E. A., Shaw, K. M., Saari, L. M., & Latham, G. P., 1981. Goal Setting and Task Performance: 1969-1980, *Psychological Bulletin*, **90**: 125-152.

Lukas, B. A, and Ferrell, O. C., 2000. The Effect Of Market Orientation On Product Innovation, *Journal of the Academy of Marketing Science*, **28(2)**: 239-247.

Lynn, G. S., 1998. New Product Team Learning: Developing and Profiting From Your Knowledge Capital, *California Management Review*, **40(4)**: 74-93.

Lynn, G. S. and Reilly, R. R., 2002. *Blockbusters: The Five Keys to Developing Great New Products*, HarperCollins Publishers, Inc. Chap. 2, pp. 15-31.

Lynn, G. S., Abel, K. D., Valentine, W. S., and Wright, R. C., 1999. Key Factors in Increasing Speed to Market and Improving New Product Success Rates, **28**: 319-326.

Mascitelli, R., 2000. From Experience: Harnessing Tacit Knowledge to Achieve Breakthrough Innovation, *Journal of Product Innovation Management*, **17(3)**: 179-193.

Millson, M. R. and Wilemon, D., 2002. The Impact of Organizational Integration and Product Development Proficiency on Market Success, *Industrial Marketing Management* **31**: 1-23.

Miner, A. S., Bassoff, P., and Moorman, C. 2001. Organizational Improvisation and Learning: A Field Study, *Administrative Science Quarterly*, **46**: 304-337.

Moorman, C. and Miner, A. S., 1998a. The Convergence of Planning and Execution: Improvisation in New Product Development, *Journal of Marketing* **62**: 1-20.

Moorman, C. and Miner, A. S. 1998b. Organizational Improvisation and Organizational Memory. Acadmey of Management: *The Academy of Management Review*, **23**: 698-723.

Nunnally, J.C. Psychometric Theory, 2nd edition. McGraw-Hill, New York, 1978.

Pearson, C. M. and Clair, J.A., 1998. Reframing Crisis Management, *Academy of Management Review*, **23(1)**: 59-76.

Pearson, C.M. and Mitroff, I. I., 1993. From Crisis Prone To Crisis Prepared: A Framework For Crisis Management, *Academy of Management Executive*, **7(1)**: 48-59.

Perry, L.T., 1991. Strategic Improvising: How to Formulate and Implement Competitive Strategies In Concert, *Organization Dynamics*, **19(4)**: 51-64.

Podsakoff, P. M., and Organ, D., 1986. Self-Reporting In Organizational Research: Problems and Prospects, *Journal of Management* **12(4)**: 531-544.

Quarantelli, E. L., 1988. Disaster Crisis Management: A Summary Of Research Findings, *Journal of Management Studies*, **25(4)**: 373-385.

Rosenthal, S.R., 1992. *Effective product design and development: How to Cut Lead Time and Increase Customer Satisfaction*, Homewood, IL: Irwin Professional Publishing.

Samra, Y.M., 2005. Is New Product Development the Solution to a Crisis or Is Crisis the Solution to New Product Development, *Doctoral Dissertation*, Stevens Institute of Technology: Hoboken.

Samra, Y.M., Lynn, G. S., and Reilly, R. R., 2008. Effect of Improvisation on Product Cycle Time and Product Success: A Study of New Product Development (NPD) Teams in the United States. *International Journal of Management*,**25(1)**: 175-185.

Schein, E.H., 1993. How Can Organizations Learn Faster? The Challenge of Entering the Green Room. *Sloan Management Review Winter*, 55-62.

Shaluf, I. M., Fakharu'l-razi, A. and Said, A. M., 2003. A Review Of Disaster and Crisis, *Disaster Prevention and Management*, **12(1)**: 24-32.

Shepherd, Charles and Ahmed, Pervaiz K., 2000. NPD Frameworks: A Holistic Examination, *European Journal of Innovation Management* **3(3)**: 100-106.

Smart, C. and Vertinsky, I., 1984. Strategy and the Environment: A Study Of Corporate Responses To Crises, *Strategic Management Journal*, **5(3)**: 199-213.

Starbuck, W.H., Greve, A. and Hedberg, B. L. T., 1978. Journal of Business Administration, **9(2)**: 112-137.

Tabachnick, B. G and Fidell, L. S., 2001. *Using Multivariate Statistics, 4th edition*. Allyn & Bacon, Chap. 13, pp. 590-595, 614.

Tjosvold, D., 1984. Effects of Crisis Orientation on Managers' Approach to Controversy in Decision Making, *Academy of Management Journal*, **27(1)**: 130-138.

Webster's Encyclopedic Dictionary, School and Office Edition, The World Publishing Company, 1993.

Weick, K. E., 1977. Organization Design: Organizations as Self-Designing Systems, *Organizational Dynamics*, **6(2)**: 31-46.

Weick, K. E., 1988. Enacted Sensemaking In Crisis Situations, *Journal of Management Studies*, **25(4)**: 305-317.

Weick, K. E., 1993. The Collapse Of Sensemaking In Organizations: The Mann Gulch Disaster, *Administrative Science Quarterly* **38**: 628-653.

Zack, M.H, 2000. Jazz Improvisation and Organizing: Once More from the Top, Organization Science, **11(2)**: 227-234.

Identifying Discontinuities for Strategic Research

Jack Hipple, Innovation-TRIZ, 18222 Collridge Dr., Tampa, FL 33647

R&D and technology planning frequently is done via extrapolation of past businesses and technologies, using the core competencies of the existing organizations. When a business is expanding along its original paths, this is a perfectly acceptable way of doing short term strategic planning. Examples of this kind of approach might be finding new ways to modify the thermal profile of a polymerization reaction to optimize yield, molecular weight distribution, or toughness structure. The switch to a catalytic system that was more specific in its targeting of molecular properties would not necessarily be obvious to one thinking only about optimizing thermal profiles. We have seen much activity in the area of very sophisticated chemical catalytic systems over the past 10 years. We now see some research into microwave catalytic systems that are not yet economic, but are certainly moving into a new "field" for providing the polymer properties that are desired. Extrapolating past technologies as the only approach to strategic research planning can have disastrous consequences when a given technology is totally displaced by a replacement technology which is totally foreign to the organization. The skill set required may be totally absent within an organization, allowing too much tunnel vision about the best way to problem solve or provide the next generation of technology. Other everyday examples include the evolution of "image capturing" from etching to painting to printing to wet chemical photography to electronic photography. Each of these approaches uses different skills and competencies and significantly different technical approaches.

When one studies the evolution of breakthrough inventions, a predictive set of technology development lines can be deduced, which allow improved strategic and R&D planning that stays one step ahead of discontinuities and anticipates them. These patterns of evolution are a key part of the TRIZ "Inventive Problem Solving" process which has been derived from the study of the global patent literature. This allows an organization to plan much more effectively in terms of budgets, types of technical disciplines needed, and types of customers and potential customers with to partner.

In addition to the field evolution line briefly mentioned above (mechanical, thermal, chemical, electronic, electromagnetic), there are others that can be deduced that are of everyday practical value. Three of these are:

1. Contradiction resolution. The evolution of products and technologies shows that systems achieve their next breakthrough by resolving contradictions in performance and design, and that breakthrough patents always resolve a major contradiction. These principles can be captured in a useful, retrievable way in a series of contradiction tables that can be used

for idea stimulation as opposed to random brainstorming. An example of this is the evolution of the automobile and its various sub-systems. When the automobile was first invented, its speed was minimal. At some point in time, probably when the car could go more than 5 MPH, the contradiction of speed vs. safety (stopping) was obvious and brakes were invented. As speed increased further, the need for smooth roads and shock absorbers became evident. Transmissions and disc brakes came along later, all in response to a contradiction in system performance. Even within these specific inventions, there are contradictions in detailed design that needed to be resolved.

In the chemical process industry, we see reaction processes changed as a function of stochiometry, equilibrium, and desired product distribution.

2. Dynamism. The study of inventions and technology evolution over time shows that all systems become more dynamic and responsive over time. Examples would include power steering and automatic transmissions in cars, chemical process control systems whose parameters are changed according to process or feed conditions. In the chemical industry we see automatic control systems that can respond to incoming raw material quality changes.

3. Field evolution. This line of development has already been briefly described above. Given the fact that this line of evolution exists, we can also ask the question about the effect of next generation fields on product quality, product purity, and process control—putting us one step ahead of competito9rs who are not aware of these lines of evolution.

This talk will review these basic patterns and demonstrate their use in strategic technology planning through further illustration of their applicability within the chemical process industries and the use of chemical products and systems.

Intellectual Property Planning for the Project Life Cycle

Overview

Often, the protection of intellectual property is accomplished by happenstance, *i.e.*, little thought is given to when and how decisions about intellectual property should be made so as to maximize the value of intellectual property and to maximize the protection from competition offered by intellectual property.

Rarely indeed are intellectual property decisions fully integrated into the overall project. Further, even when addressed, the decision of when to seek protection and what protection to seek does not include the participation of critical stakeholders. The following are three typical examples:

1. *Failure to integrate R&D and the business group:* R&D develops a better mousetrap. The idea is judged, implicitly or explicitly, against other ideas developed by the same R&D group to determine whether it is worthy of spending the precious few dollars allocated to the R&D group patent budget. No one checks with the business group to determine what protection is needed in the marketplace. A patent issues, but without the input of the business group, the patent offers no real marketplace protection.

2. *Last second decisions:* After spending $15 million on the research, pilot plant, design, and part installation of the facility to manufacture the better mousetrap, a member of the project team, during a routine project meeting, asks if anyone "has consulted legal." Frantic calls are made to outside counsel who is allotted two weeks for a freedom-to-operate study, "keeping in mind that we already have this thing half-built." The freedom-to-operate study may lead to a conclusion requiring termination of the project, or, the last-second nature of the freedom-to-operate study may minimize the protection offered by the study.

3. *Failure to review the intellectual property of others* Three-quarters of the way through construction, the marketing department develops a campaign for the newest mousetrap. No one thinks to run a knockout study and within two weeks after the first product is sold, the company receives a cease and desist letter notifying it that it has infringed on the federally-protected trademark by a tiny mousetrap manufacturer based in Canada.

Unfortunately, these scenarios are all too common. Even when the protection of a company's intellectual property is systematically dealt with, the evaluation of it is often done incorrectly. Rarely is the question of "Is this a worthwhile idea to protect" compared with "What advantage will I obtain by protecting this property?"

This paper examines an alternative method to identify and protect intellectual property in conjunction with the product lifecycle. Specifically, it examines when during the project life cycle that a company should evaluate a project for potential intellectual property, and when during the project life cycle it should compare its activities with the intellectual property of others. Further, it addresses the need to include all of the critical stakeholders in the process of seeking protection of intellectual property, not just the developer of the technology. This paper focuses primarily on patents and trade secrets, but does include some discussion of trademark and copyright protection as they affect mostly-technological innovation.

Intellectual Property Protection

Intellectual property is extremely expensive property to both develop and maintain. For instance, a single United States patent will result in expenses of up to $40,000 during its lifetime and a decision to seek protection in foreign jurisdictions can result in additional costs of more than $100,000. Only a fraction of these costs are for the development of the property, *i.e.*, the actual drafting and prosecution of the patent– far more is spent on filing fees, maintenance fees, and translation fees. The decision on whether to seek protection for the potential property is not best made by the developer alone. This person or group may have little idea of the potential market at the onset of the process of development of the property.

Intellectual property takes many different forms and its development often involves complicated decision-making. For instance, should the company seek copyright or patent protection for its proprietary software, or perhaps some combination of both? While the new process used to manufacture the company's plastic is certainly new and innovative, does it make sense to patent it, or simply maintain it as a trade secret? Is it worth the time and trouble to seek federal registration of the company's long-used trade mark (and risk and Opposition proceeding), or would it be better to just stand on common law rights? Each of these approaches has benefit and risk, and such decisions may change throughout the project life cycle.

It is critical that any intellectual property development evaluative process include the input of all critical stakeholders, most often, representatives from the developers, the business unit, and the legal department. With regard to a patent, the developer most often can contribute a sense of the technological benefits of his invention and how that invention differs from what has previously been discovered or developed.

The importance of the business representative cannot be underestimated. It is the role of the business representative to contribute at least the following:

1. How this invention fits with the existing product line(s) offered by the company;
2. The market potential for this invention;
3. The legal jurisdictions where markets exist for this invention;
4. The likely competitors and those competitors' potential uses of the technology;
5. The ability of the business to exploit the invention, including the availability of both capital and people resources; and, most importantly,
6. The intended use of the property.

The legal representative's responsibility is to contribute information related to the legal effect of the property, offer options to strengthen the property, set timetables for completion, provide advice about alternatives, structure the development of the property, provide cost versus benefit advice, and obtain assistance where required for development. Further, as it is most often it is the responsibility of the legal representative to maintain the property, the legal representative should provide information about long term costs and legal requirements for property maintenance.

It is only with the contribution of at least these three stakeholders that a decision regarding the development of intellectual property can be made. Others may also be necessary, depending on particular projects. For instance, complex manufacturing methods may require process personnel, multi-business unit development may require corporate representative assistance, and trade secrets may require changes to labor and employment contracts and practices, necessitating human resource department input. It is when these stakeholders are not included in the decision-making process that the value of the intellectual property is decreased.

The timing of when to seek intellectual property protection is also critical. One illustrative example is the development of a trade secret. Often, a trade secret is not identified by a company until the secret walks out the door with a departing employee. Without the proper safeguards, that trade secret is not protectable and a company may find its most valuable alleged secrets now shared with its chief rivals because it (1) did not identify them early enough and (2) did not protect them in a way required by law. Proper management of the identification of intellectual property and development of the intellectual property at all stages of the project life cycle will generate not only protectable intellectual property, but intellectual property of the proper breadth and depth to be valuable.

A company should also decide whether it is willing to litigate to protect its intellectual property. Litigating intellectual property issues can be very expensive, but the lengths a company is willing to go to protect that property directly affects its value. How valuable is a patent when the company's competitors are fully aware that it will never seek to stop infringers? Why seek a license when you can simply infringe with impunity? On the other hand, a rabid litigation scheme will be expensive and could cost the company more in fees than the value of the property. It is critical when deciding whether and when to seek protection to determine how willing you are to protect your property.

Protection Against the Intellectual Property of Others

The most frustrating aspect of intellectual property is that it is owned not only by the company, but also by the company's competitors. A company must be vigilant to develop its own property, but must also avoid the intellectual property of others. While there have been recent significant changes in the law regarding the obligation to avoid the intellectual property of others, it still makes little economic sense to develop technology for which the company will be later estopped from using. On the other hand, it also makes little economic sense to employ high-cost legal resources too early in the project life cycle, as it may be necessary to kill the project based on other concerns. Further, there may be certain legal ramifications to identifying the intellectual property of others and then failing to follow-up on that identification with a comparison of that property to the activities of the company. The paper discusses when during the project life cycle the consideration of the intellectual property of others should be evaluated.

Input into the Project Life Cycle

Once intellectual property of the developer and competitors is identified and analyzed, this information can be included in the overall decision-making aspects of the project life-cycle. Rather than treating intellectual property as an afterthought, or simply including it too late or early in the cycle, the examination of intellectual property should be treated just as decisions on sizing, or facility location, or vendor-sourcing. It is only through proper treatment throughout the project life cycle that the overall project can be fully enhanced by intellectual property protection.

298d Secrets of Supercoaching Inventors

Gregory Leman, PhD, Department of Management & Entrepreneurship, Baylor University, One Bear Place, Box #98006, Waco, TX 76798-8006

The SuperCoach(tm) Entrepreneurial Training system has been built and optimized specifically to enable technical inventors in the discovery of exactly how and where their innovations can produce high value that can be captured. The tools and techniques used in this system drive the development of the understanding that so often highly technical would-be entrepreneurs lack when they "pitch" their opportunities to angel or venture capital investors, and can be equally overlooked in the corporate settings where technical and business leaders remain separated into functional silos. A key insight from combined experience of several top entrepreneurial coaches can be summarized by the statement, "No one writes their way to success - but they must talk their way" demands that a cohesive story of how the business will actually create value and how that value will be defended. The process of how that story is most efficiently created, and its assertions defended are reviewed and exemplified. This methodology is contrasted and compared to classic models of mentoring and consulting.

298e Managing Technological Innovations: Tips for Scientists and Engineers

Mukund Karanjikar, Chevron Energy Technology Co, 3901 Briarpark Drive, Houston, TX 77042

The lack of innovations is often blamed on management policies or on the gap between 'technical' and 'commercial side' with their divergent perspectives. Many times however the owner of the idea (i.e. engineer, scientist or technologist) can do a number of additional activities in order to make his or her idea a 'package' that can be sold to both insiders and outsiders.

This presentation discusses ten tips for innovators in order to enhance their chances of success. It is based upon an article published by the author in November issue of Chemical Engineering Progress (1).

References:

1. M. Karanjikar, 'Corporate Venturing and Pursuit of Innovation: Ten Tips for Intrapreneurs', Chemical Engineering Progress, Forthcoming, November 2007, pp 45-47

298f The Changing Role of Patent Opinions In Product Development

Neal Seth, Baker Hostetler LLP, Washingon Square, Suite 1100, 1050 Connecticut Ave. NW, Washington, DC 20036

Recent modifications in the law have changed the effect of legal opinions directed to patent clearance/ non-infringement that were previously sought before developing or launching a new product. Learn how the new standard should shape the type of patent legal clearance needed and sought.

479a Bioenergy Development through University Research Centers

Nemoy Rau, Department of Chemical Engineering, Tennessee Tech University, Box 5013, Cookeville, TN 38505 and Hans Rau, Finance, The Kazakhstan Institute of Management, Economics and Strategic Research, Economics and Strategic Research (KIMEP), 2 Abai Avenue, Almaty, Kazakhstan.

The search for the renewable, alternative fuels has become critically important in the last couple of years with the crude oil prices increasing beyond $100 per barrel with no decline in sight. Many universities are increasing their resources to bioenergy by establishing research centers to bring together the interdisciplinary research required in bioenergy between various academic disciplines. With the anticipated increases in federal funding to bioenergy in the next five years, more universities are planning to establish Bioenergy Research Centers. There has been no consolidated research to catalog and evaluate the accomplishments of the bioenergy research centers hitherto. This study will identify the existing Bioenergy Research Centers and their work at the Tier 1 and Tier 2 universities in the United States and Canada. In this study, we will compare the research focus of these research centers and analyze their research, collaborations, organization structures, areas of specialization and their success in fulfilling their stated mission. A survey instrument will be designed to collect specific information on state and federal funding received, number of researchers, involvement with graduate and doctoral programs, and their future strategies. We expect this study to provide consolidated comparative information on the common areas of research pursued, areas with lesser focus, interaction with the graduate curriculums. The major accomplishments of this study will be a summarization of the current university Bioenergy Research Centers' fields of concentration and comparative evaluation of their accomplishments. We will report on the characteristics and structure of the more successful centers and look at the collaborations that focus on the regional specific needs and future economic development goals.

479b Necessary Preliminaries to Innovation In Univerities

Richard L. Long, New Mexico State University, Chemical Engineering Department, P.O. Box 30001, MSC 3805, Las Cruces, NM 88003-8001

Universities are being driven in part by some of the same market forces that are driving business. If one listens to the rhetoric of university presidents it sounds much the same regardless of its source. The drive for technological and other types of innovations is the drive for funding. This paricularly affects units with the potential prospect of obtaining large amounts of primarily government funding because this is the source of so-called "overhead" dollars which puts financial discretion decision-making power in the hands of various administrative officials. That discretion can be used for many development purposes.

479c The Institute for Collaborative Biotechnologies

David H. Gay, Institute for Collaborative Biotechnologies, University of California, Santa Barbara, Santa Barbara, CA 93106-5100

The ICB is a uniquely interdisciplinary alliance of academia, industry and the Army. Led by the University of California, Santa Barbara, in collaboration with MIT, Caltech, industry partners and the Army, the ICB transforms biological inspiration into technological innovation.

Inspired by the study of the amazing processes, structures and features found in nature and biology, ICB research teams develop revolutionary technological innovations in: bio-inspired materials and lightweight portable energy; biomolecular sensors; bio-inspired energy-dispersive composites; biodiscovery tools; bio-inspired network science; and cognitive neuroscience.

The ICB is by name and by design an intensely collaborative endeavor that is focused on technology transition in close partnership with Army and industry researchers. The presentation will discuss the challenges encountered, the synergies realized, and the lessons learned during the first five years of operation of the Institute for Collaborative Biotechnologies.

Web Page: www.icb.ucsb.edu

479d A New Government, Industry and University Partnership through Discovery Science and Engineering Innovation Institutes

Justine E. Johannes, Materials Synthesis and Processing Department, Sandia National Laboratories, P. O. Box 5800, MS 0885, Albuquerque, NM 87185-0885, Duane Dimos, Material Science and Engineering, Sandia National Laboratories, P.O. Box 5800, MS 0885, Albuquerque, NM 87185-0885, Regan Stinnett, National Institute of Nano Engineering Program Manager, Sandia National Laboratories, P.O. Box 5800, MS 0885, Albuquerque, NM 87185-0885, and Paul Smith, Licensing &IP Management, Sandia National Laboratories, P.O. Box 5800, MS 0885, Albuquerque, NM 87185-0885.

America's approach to technical innovation must undergo transformation for the country to successfully compete in the global technology race. According to the National Academies, Rising Above the Gathers Storm report, the following actions must occur for a successful transformation: increasing the talent pool through improved K-12 education, sowing the seeds through a strengthened commitment to long-term research, attracting and retaining the top scientist and engineers, and developing a culture and environment to encourage innovation. In an effort to accelerate this transformation, the President introduced the America Competitiveness Initiative (ACI), a bold action to increase national investments in research and development, math and science education and innovation/entrepreneurship. In this presentation we will describe efforts underway at the Sandia National Laboratories to stand-up Discovery Science and Engineering Innovation Institutes (DSEII) as part of the ACI.

DSEIIs will be technology incubators and centers of learning, for undergraduate and graduate students, that build off the expertise and facilities resident at the host national laboratory and will bring together academia, industry and government in powerful partnerships. The network of National Laboratories, and the investments they represent, are significant and can provide a unique opportunity to accelerate the translation from idea to innovation. Developing a system of DSEIIs focused on key research areas and critical technologies will facilitate cooperative innovation and offering new educational (graduate and undergraduate) opportunities with a national reach and a strong regional impact.

The prototype DSEII currently being proposed by Sandia National Laboratories and their partners is the National Institute for Nano Engineering (NINE). Nano-engineering is one of the coming century's most exciting and potentially important areas of science and technology and is one whose economic benefits will depend on an educational infrastructure that supports the training of students, teachers, university faculty, research and technical work from a diverse set of technical disciplines. NINE represents a new paradigm of partnership among universities, industry and national laboratories. The Department of Energy and the National Nuclear Security Agency have made significant investments at Sandia National Laboratories in microsystems, nano-sciences and high-performance computing which provide key capabilities needed to drive future competitiveness in nano-scale technologies and transform education in the field of nano-engineering. NINE will be a distributed enterprise, with its hub at Sandia and with spokes at major universities and with key industrial participants. NINE would demonstrate the path for creating new entities which are both technology incubators and centers of learning. An understanding of how NINE will operate and the unique partnership being established with the partners will be discussed.

Web Page: www.sandia.gov/NINE/

IP Issues in Worldwide Technology Collaborations

by Tom Adolph
Jackson Walker LLP
713-752-4208
tadolph@jw.com

As the world shrinks, and as more and more companies are involved daily in technology and other intellectual property collaborations, the complexity of issues that arise in these collaborations increases. The issues range from the mundane to the surprising to those that require deep thought and complex solutions. Because of the large number of jurisdictions and the infinite amount of possible collaborations, it is impossible to address all issues that may arise. This paper will simply attempt to address some common issues.

License language

The license grant provision is usually the most important language in any license. The license grant should track the language of statutes that define the intellectual property rights at issue. For example, in the U.S., a patent grants the rights to exclude others from making, using, selling and importing the claimed invention. Other US IP rights use different language to describe the exclusive rights, and certainly statutes in other countries use their own language to define the IP rights. It is crucial that any multi-country license use the appropriate language for the IP rights included in the license.

In worldwide collaborations, translation issues can become critical. The implications of a single word are important in any license grant, and translation issues may make it difficult to discover misunderstandings about what rights have been granted until late in the relationship. A grant "under a patent," for example, is ambiguous. Does it include all rights under the patent? The right to sublicense or assign? The right to make?

See Appendix A.

Licensed rights ownership – due diligence

Before any company commits to a license in which it is to be a licensee, it is crucial that the company perform proper due diligence to determine whether the proposed licensor truly has the necessary rights in the property to be licensed. In dealings with foreign licensors, this can be a very difficult task. The task is even more difficult when the licensor is a foreign government or when the licensor derives its own rights from a foreign government. Due diligence often reveals that the promised rights are non-existent, or are fatally conditional. Often, the licensor and the foreign government may be willing to correct problems disclosed by the due diligence, but the process of correction can be painful and time-consuming.

See Appendix B

Licensed rights – territorial issues

Territorial issues arise in many ways in multi-country IP collaborations. One such issue concerns the territorial limits of patents and trademarks. If a company obtains a worldwide license from a company under a single patent or trademark from a single country, the "world-wide rights" may be ethereal. Due diligence is required to understand the true territorial extent and limits of the IP rights to be licensed. Once the territorial limits are understood, then analysis is required to determine whether the IP or the technology has any real value or potential value beyond those territorial limits. Technology patented in only limited countries may still have value as a trade secret in other countries, or, conversely, the patents may have destroyed the trade secret value of the technology. Trademarks protected only in certain countries may have no value outside those limits. Squatters may have already staked out trademark protection outside the limits of the licensed rights. It is crucial that one understand the true territorial limits of the licensed rights and the implications of those limitations.

See Appendix C

Licensed rights – governmental rights and regulation issues

Multi-country technology or IP collaborations often run into issues where local governments have, by law, rights in the licensed technology or IP. Other times, a government may, after the license is executed, declare that it has rights in the IP or technology. Still other issues arise simply because a local government has regulated the IP or technology and requires certain licenses before certain events under the license may occur. Bureaucratic nightmares often interfere with multi-country collaborations, causing what seemed like an event that would occur quickly to become a far off goal that cannot be reached because of bureaucratic red tape. All of these issues may be unfortunate surprises after-the-fact. Again, due diligence is required to avoid such surprises.

See Appendix D.

Sovereignty and governmental pride

When more than one country is involved, what might be deemed "governmental pride" can become an issue. Most government contracts require dispute resolution to occur within their borders. When two or more governments are involved in a collaboration, this can become a Catch 22. Government A will not agree that the law of Government B will govern the contract or that arbitration or lawsuits will occur on Government B soil. Truly creative solutions are often called for to resolve these issues.

Developed IP Ownership

In any license of IP or technology, the issue of who owns technology or IP developed during the term of the license is important to all parties. Pride, mistrust, suspicion, secrecy, and government requirements all play a part in resolving these issues and in living with the resolution.

Entity ownership

In multi-country collaborations, national laws often dictate the creation of a new legal entity as the licensee – an entity owned by multiple parties and often even in part by a government. Often, national constitutional issues come into play in this area. Requirements for specific minimum percentage ownership by nationals or for maximum ownership percentages by foreigners, have implications for how the licensor may be willing to license the IP or technology. This is yet another area where creativity is often required.

Employee Secrecy and IP Ownership Issues

US companies are often surprised to learn that despite their best efforts to protect their IP from misappropriation by their international partner, the employees of the international partner do not share in the obligations. One must insure that the employees understand their obligations and agree to them. Obligating the international partner to enforce the obligation may also be helpful.

Forced licenses of IP, Forced IP transfer, Restricted Terminations

Some countries have legal requirements that force IP or technology grants to national entities, or force assignment of ownership in IP or technology, or that severely restrict a foreign licensor's ability to terminate a licensee. An owner of valuable technology or IP must be extremely careful not to inadvertently fall into one of these legal traps. Doing business in such countries is still possible, but it requires due diligence and careful deal structuring.

Real world issue – exporting or importing products, technology, IP, and money

Multi-country collaborations often run into complications involving the export or import of products, technology, and IP, or even in just getting money in and out of the country. Many products and much technology is heavily regulated or restricted with respect to import or export. Issues of national security often come into play. Further, many countries restrict the ability of a foreign company to take money out of the country. These issues often require bureaucratic solutions and creative legal relationships. For example, Brazil requires registration of licenses. Renewal of the license requires a showing of new improvements. The licensor must also demonstrate actual technology transfer.

Litigation issues – Can I enforce my rights?

As with any other legal relationship, in a multi-country collaboration, one must think ahead to the possibility of a problem. What happens if the deal goes bad? How can I enforce my rights? This issue becomes very complicated on the world-wide scale. For example, it is illegal in some countries simply to take a deposition. It may be impossible ever to get the testimony or other evidence needed to prove you case. Other countries are notorious for their lack of IP enforcement. Any multi-country collaboration must consider whether the deal can ever really be enforced or whether all depends on the trustworthiness of the other party.

Antitrust issues in licenses

All collaborations of any kind – within a single country or across several countries, can have antitrust implications. Collaboration – at least improper collaboration – is the basis of much antitrust law. Due diligence is required to understand the antitrust implications of any such collaboration. Among other possible areas, antitrust issues can arise with respect to price provisions, non-compete provisions, territorial restrictions, and termination rights.

Warranty and warranty remedy issues

In any licensing arrangement, the agreements must be clear about all warranties, all exclusions of warranties, and all remedies for warranties. In any multi-country transactions, the exclusion of warranties is critical and sensitive. A party must be clear about all warranties that it wishes to exclude. But problems can arise if the party is unaware of implied warranties that must be expressly mentioned in the exclusion or worse still, if the party expressly excludes a warranty that the law prohibits a party from excluding. As always, due diligence is required.

IP infringement suits

In today's world of patent troll suits, any IP or technology collaboration must consider the possibility of patent suits against one party or the other. Conversely, the parties should also agree how to handle IP suits against others relating to the collaboration. Some of the most difficult issues concern who will pay for and manage the suits and how to split the damages awarded.

IP maintenance fees

A somewhat mundane issue, but a very important issue, concerns maintenance fees for IP. For example, most patents and trademarks require periodic payment of "maintenance fees" to continue or renew the patent or trademark. Failure to pay the fees leads to abandonment of the property. Usually, the licensor in any license will retain this right. But the licensee should have failsafe provisions. It is often the licensee who has the biggest stake in the continuing viability of the patent or trademark. Therefore, the licensee will want to obligate the licensor to pay the fees and the licensee will further want to have the right to pay the fees if the licensor fails to pay or fails to provide assurance of payment.

Grant back issues and grant back extension issues

Many multi-country IP or technology collaborations have grantback licenses that may be simple or may be complex. Some grantback licensing arrangements involve networks of licensees and sublicensees. The grantbacks can be valuable for all parties, but they can also be a nuisance, or worse still, can become an attack on the "family jewels" of a company. Careful negotiations are required to strike the proper balance.

Appendix A
License Grant Language

Example 1

Licensor grants to Licensee a fully-paid-up, exclusive license under the Patent:

A. to use, sell, import, transfer, and distribute the Products in the Territory, and to grant sublicenses under all such rights; and

B. upon occurrence of an Escrow Event, to make, have made, use, sell, import, transfer, and distribute, modify, copy, and create derivative works of the Products in the Territory, and to grant sublicenses under all such rights, all solely for the following purposes and for no other purpose:

1. to complete and deliver (and exercise other necessary rights relating to) Products which, as of the occurrence of the Escrow Event:
   (a) were already ordered, or
   (b) were in progress of being assembled, or
   (c) were the subject of an existing contractual obligation of PI to provide to any person, or

2. to repair, maintain and support (and exercise other necessary rights relating to) existing Products or Products completed as permitted above.

Example 2

Subject to the terms and conditions of this Agreement, Licensor grants to Licensee a nonexclusive, royalty-bearing, terminable, license under the Patents and under all trade secret rights and other intellectual property rights of Licensor relating to the Technology (but not trademark rights or trade name rights) to make (but not to have made), to use, and to sell Licensed Products. Licensee shall have no right to grant sublicenses.

Example 3

Licensor grants to Licensee a royalty-bearing license under Licensed Patents to manufacture, have manufactured, use, offer for sale, sell, lease, and/or import Licensed Products in the Licensed Field in the Licensed Territory. This license grant is exclusive only to the extent indicated in Section 1 of the Patent License Agreement. Otherwise, the license grant is non-exclusive.

Appendix B
IP Due Diligence

Schedule all domestic and foreign patents and patent applications of the Company, and get copies of all issued, unexpired patents.

Schedule of all domestic and foreign trademarks, trade names, service marks, brand names, logos and other trade designations (including unregistered names and marks) and trademark and service mark registrations and applications (state and federal) of the Company.

Schedule all domestic and foreign copyright registrations, and applications to register copyrights of the Company.

Prepare a chain of title of each item of Intellectual Property, together with all documents that disclose, evidence or refer to any part of such chain of title.

Obtain copies of all agreements with third parties for the design and development of the Company's products.

Appendix C
Territory Issues

Example 1

"Licensed Territory" means the territory so indicated as the Licensed Territory in Section 1 of the Patent License Agreement, subject to the then-current applicable article, item, service, technology, and technical data-specific requirements of the U.S. export laws and regulations.

Licensee will use diligent and commercially reasonable efforts to Actively Commercialize Licensed Products in the Licensed Field in the Licensed Territory.

Licensee will pay a running royalty equal to the percent of Net Sales for Licensed Products Sold in each Contract Quarter.

Example 2

The "Territory" means the United States of America, including all territories, districts, and possessions.

Appendix D
Governmental Rights and Regulation

Example 1

Compliance: At its own cost, Licensee shall comply with all rules and regulations of every administrative or other governmental authority located in the Territory having jurisdiction over the handling, use, or sale of the Products. Licensor shall, at its cost, make any changes or modifications to the Products required for the Products to comply with all rules and regulations of every administrative or other governmental authority located in the Territory having jurisdiction over the handling, use, or sale of the Products. Licensee, with the reasonable cooperation of Licensor, shall comply with applicable U.S. Food and Drug Administration laws and regulations. Licensor shall work with Licensee to meet all US statutory and regulatory requirements and to meet the requirements of Licensee's customers. If the cost to Licensor of changes or modifications required under this section exceeds $xxx, then the parties shall, in good faith, attempt to agree to a method of shared costs.

Regulatory Assistance: Each party shall cooperate with the other in any efforts to meet regulatory and other governmental approvals and requirements.

535a Transitions to Management: Testing the Water Before Jumping in the Deep End

Shawna S. Berg, Process Engineering and Evaluations, Shell Global Solutions (US) Inc., Westhollow Technology Center, P. O. Box 1380, Houston, TX 77251

Have you had experience with poor management and feared making the same mistakes as a future manager? Have you ever wondered if you had management potential? Many engineers will make the transition from traditional engineer to manager at some point in their career. Unfortunately, this transition often comes with little or no formal guidance as to how to be a good manager. This paper discusses the skills and awareness needed to take the leap into management as well as the ways to test the management waters early in a career before becoming a manager.

535b How to Be a Successful Manager In a Technical Organization

Jonathan Worstell, Hoder, Shell Global Solutions US, 3333 Highway 6 South, Room CP-109, Houston, TX 77082

Until the 1990's, an engineer or scientist aspiring to be a manager in a technical organization first had to prove their technical competency, and second, that they were technically the most qualified for such a position. With the advent of the MBA, technical organizations no longer require technical competency from managers, particularly "first line" managers.

This presentation discusses the impact of the MBA on technical management and reviews the careers of nationally recognized technical managers. This presentation questions the need to have an MBA to grow into a management position and asserts that such a "need" adversely impacts technical organizations.

A Career in Project Management!!

Stephen L. Cabano, Pathfinder, LLC, Cherry Hill, NJ USA

Introduction

Selecting a career is certainly one of the most significant decisions one has to make in his/her lifetime. With so many interesting, varied, challenging and rewarding occupations to choose from in today's diverse and dynamic world, this is truly a perplexing dilemma. I'd like to share with you my perspective as an admitted advocate and long time practitioner for considering the challenging but enormous vistas of gratifying experiences, opportunities and satisfactions available for selecting project management as a career vocation.

This paper/presentation combines the perspective of my 25 years in project management with that of my father, who has been a career professional in the domestic and international Petroleum, Petrochemical and Chemical industry. Hopefully this point of view will give an insightful look at the diversity, excitement, technical, financial, ethical & cultural challenges and holistic combination of experiences and issues that all interact in most Project Management situations, to make Project Management the ultimate field of endeavor for some, but certainly not all.

Before launching into the various aspects of a career in Project Management, let's try to establish a couple of basic definitions.

What is a Project?

"A one-time, specific task to be accomplished within prescribed quality and economic parameters in a finite time period. Task should include a definition of desired end results in terms of a facilities and/or work description, containing measurable performance criteria."

What is Project Management?

"The function of identifying/planning, organizing, resourcing/staffing, implementing/directing and controlling an array of often diverse activities in a unified effort to perform a defined task within pre-established performance/achievement parameters."

Diversity of Project Environment Experiences & Challenges

Let's recognize right up front that we are all unique individuals, each having our own collection of likes and dislikes, as well as competencies and limitations. No profession will fit or satisfy all or even any large group of us; so a discussion of this type should start out with a listing of the general characteristics, interests and attributes that most successful Project Managers have in common.

Let's first start off by discussing some of the framing characteristics of a typical project environment. Project evolution can be staged in numerous ways; however, one of the most popular uses five sequential phases:

1. Business Planning or Defining the Opportunity!
2. Analyzing Alternatives!
3. Defining the Selected Option!
4. Executing the Selected Option!
5. Operate & Evaluate!

The interface between each of these phases is an important evaluation and decision point. The thoroughness of conducting each phase and the attractiveness of the project at each juncture helps determine whether the project should move forward into the next phase, recycle back for improvement or be terminated.

Capital projects often involve a diverse mix of business/financial, scientific/technological, human/social/cultural issues with relatively large groups of people with different skill sets and competencies, frequently in different locations, all trying to develop/implement a synchronized an efficient, competitive solution for the project sponsor. Throughout this process, there are frequent & numerous temptations and pressures to digress from proven "best practices" and carefully designed plans usually in attempts to cut cost and/or accelerate schedule – and do all this while trying to earn a reasonable return for their expertise and respective contributions.

Furthermore, the playing field is often in a state of flux, with business, technological, social, environmental, marketplace, resource and political, financial situations often changing concurrent with the development and various phases of the project. Thus, inherent in this dynamic environment is the ever-present need for a stabilizing, rational influence, excellent communications, along with risk awareness and adroit management. Obviously, teamwork and common focus are also vital throughout. Above all is the recognition that "People Build Projects, Not Companies" and that "Better People Build Better Projects."

Most Significant Attributes/Requisites for Success

Having set the Project Management backdrop, let's now look at some of the more important attributes and requisites for those who enjoy the most fulfillment and success in this field. Obviously, it's not required to be a master of all these characteristics, but certainly a candidate should be proficient in many and feel comfortable in essentially all. I have attempted to separate the more significant of these into an initial major group:

- Holistic vision and perspective.
- Understanding of and respect for the varied and complex talents, skills and competencies and how they interface/interact to successfully carry out an industrial capital investment project.
- Inclination to organize and facilitate groups of diverse subject matter personnel to analyze problems, develop solutions and implement same – get things done!

- Diverse experience; understands/respects varied & complex contributions/facets of a project.
- Create and maintain an environment that allows knowledgeable people to work enthusiastically, creatively, efficiently and effectively.
- Compulsion for organizing, planning and executing creative initiatives with shared participant satisfaction.
- Technical, business, legal and financial appreciation.
- Ability to distinguish between relevant and irrelevant.

In addition, there are a number of other attributes/requisites which are also desirable:

- Ability to think both strategically and tactically.
- Good human interaction skills – diplomatic but precise.
- Practical/sound balanced judgment.
- Resistant to stress and strain.
- Organizationally/politically astute.
- Open-minded and decisive.
- Sets and monitors objectives.
- Determined, dedicated and dynamic.
- Organized and an organizer.
- Solid contract administration skills.
- Achievement oriented.
- Respect for people and their sensitivities.
- Persuasive and articulate – communicator.
- Inspires confidence and team work.
- A team "leader" or team "player".
- People development interests.
- Organized and predictable.
- Persuasive/convincing.
- Strong facilitation skills.
- Determined but realistic.
- Creative, flexible but focused.
- Like to work in a group, rather than alone.

Rewards, Fulfillment & Gratification Potential

Now let's turn to some of the benefits of a career in Project Management for those who have the interest, aptitude and capabilities. I can personally attest to the reward, fulfillment and gratification potential for a career in this vocation. For most of my life I've been around people immersed in the project business and have observed and experienced the satisfactions as well as the disappointments involved. From my perspective, the positive aspects significantly outweigh the negative. Among the many significant benefits offered include:

- Pride & accomplishment.
- Responsibility/accountability oriented.
- Combination of technical & managerial skills & leadership.
- Tangible, visible results.
- Team camaraderie, ever expanding life-long friendships.
- Exciting, diverse, constantly challenging work environment.
- Both personal and team satisfaction.
- Collaborative, unified work setting.
- Interesting, progressive, learning experiences.
- Psychic income.
- Well paying and prominent.
- Job recognition, respect & security.
- Multi-faceted and multi-discipline situation mix.
- Competitive.
- Interesting and gratifying.
- Contributing and value-added.
- Dynamic ambiance.
- Compliant yet innovative.
- Personnel development.
- International exposure.

Since projects are a micro mix of the variety of challenges and situations we usually encounter in every day life, we should expect that they would offer a wide range of interesting and gratifying opportunities. Obviously, it's not everyone's elixir, but if it does suit your taste and talent, it might well be your path to success and satisfaction.

Limitations & Sacrifices

As with any profession, there are certainly some downside features and Project Management is no exception. Among these, one might include:

- Lots of travel – both domestic & international!
- Usually somewhat longer work hours than involved discipline personnel.
- Broader, more diverse background and experience requisites.
- Collaborative rather than self-dependency.
- High stress with pressure to get the job done.
- Greater visibility.
- More responsibility & accountability.
- Less technical and more managerial/impersonal skill driven.

Opportunities Available – Welcome Aboard

Clearly there are certain attributes, interest areas and skill sets that, if properly combined and nurtured, provide the right foundation for a very inspiring and gratifying career in this important and valued field.

You may have noticed that you can hardly pick up a technical or business magazine or journal these days that doesn't advocate the benefits of **Project Management** to corporate development and progress. Project Managers are being sought to lead many corporate initiatives with a much broader charter than ever before. Leading Project Managers are being looked upon as "Venture Managers" to take the lead in pursuing new research and development efforts, development of international operating and manufacturing centers, acquiring new technology and assets, etc. You name it and it's a potential opportunity for someone with Project Management skills and proficiency. Most leading corporations have clearly identified Project Management as a core-competency. In fact, many corporations are increasing their Project Management staffs and reducing other in-house skills which can be outsourced and assigning management of these outsourced functions to their Project Management personnel. The field of Project Management continues to expand in almost every direction, and is only limited by the availability of people with the interest, skills, dedication and commitment to fulfill these expanding horizons.

This article addresses the personal attributes and areas of interest, which combined, convert what would otherwise be a routine daily job into an exciting, interesting, challenging, fulfilling and thoroughly enjoyable and value-added lifetime crusade for some, while earning an impressive income to boot. A Project Management career is not suitable for all, but for those with the right credentials and interests, it can be the "wind beneath your wings" or, for some of us, "what lights your Bunsen burner."

PlantTriage

Delivering Business Results

from Engineering to Management

George Buckbee, P.E.

ExperTune®

PlantTriage

Agenda

- What Engineers Want
- What Managers Want
- Tools & Techniques
- Case Studies
- Q&A

PlantTriage

What Engineers Want

In the day-to-day work environment:

- WHO do you have lunch with?
- WHAT do you work on?
- WHEN do you report results?
- WHERE do you spend any free time?
- WHY are you working on your favorite project?
- HOW do you know you did well?

PlantTriage

What Managers Want

In the day-to-day work environment:

- WHO do they have lunch with?
- WHAT do they work on?
- WHEN do they report results?
- WHERE do they spend any free time?
- WHY are they working on their favorite project?
- HOW do they know **<u>you</u>** did well?

PlantTriage

The GAP

ENGINEERS

- Understand the details
- Technical Results
- Metrics adapt to the problem
- Enjoy solving the problem

MANAGERS

- Don't care about the details
- Monetary Results
- Fixed set of metrics & reporting method
- Enjoy communicating the results!

PlantTriage

Tip#1

- Work <u>ONLY</u> on the right things.

PlantTriage

Calculate the Value

- Cost
- Quality
- Safety
- Reliability
- Production
- Energy

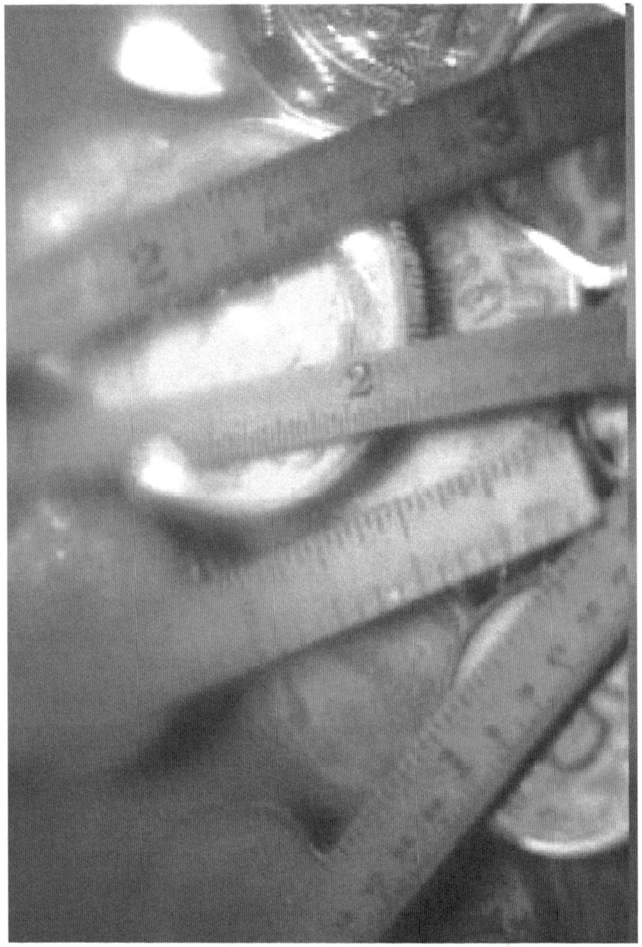

PlantTriage

How?

- Before & After
- Use Real Data, Real-time
- Estimate
- Simulation

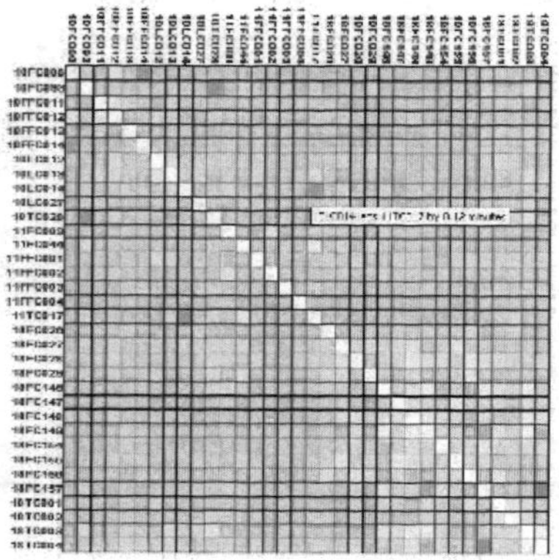

PlantTriage

The Most Important Technique

PlantTriage

Communication Techniques

- Before & After
- 1-Page Summary
- Elevator Speech

- And **NOT:**
  - Detailed Report

PlantTriage

Case Study #1

PlantTriage

Case Study #2

PlantTriage

Q&A

AUTHOR INDEX